The New Fresh Seafood Buyer's Guide

A manual for distributors, restaurants and retailers

The New Fresh Seafood Buyer's Guide

A manual for distributors, restaurants and retailers

Ian Doré

An Osprey Book
Published by Van Nostrand Reinhold
New York

An Osprey Book
(Osprey is an imprint of Van Nostrand Reinhold)
Copyright © 1991 by Ian Doré

Library of Congress Catalog Card Number 91-11757
ISBN 0-442-00201-7

All rights reserved. No part of this work covered by the copyright hereon may be reproduced or used in any form or by any means—graphic, electronic, or mechanical, including photocopying, recording, taping, or information storage and retrieval systems—without written permission of the publisher.

Printed in the United States of America.

Van Nostrand Reinhold
115 Fifth Avenue
New York, New York 10003

Chapman and Hall
2–6 Boundary Row
London, SE1 8HN, England

Thomas Nelson Australia
102 Dodds Street
South Melbourne 3205
Victoria, Australia

Nelson Canada
1120 Birchmount Road
Scarborough, Ontario M1K 5G4, Canada

16 15 14 13 12 11 10 9 8 7 6 5 4 3 2 1

Library of Congress Cataloging-in-Publication Data

Doré, Ian, 1941–
 The new fresh seafood buyer's guide : a manual for distributors, restaurants, and retailers / by Ian Doré.
  p. cm.
 Includes bibliographical references and index.
 ISBN 0-442-00201-7
 1. Seafood—Purchasing. I. Title.
TX385.D68   1991
641.3'92'0687—dc20   91-11757
                                             CIP

Preface

This book is a completely new edition of *Fresh Seafood—The Commercial Buyer's Guide*, which was first published in 1984. There have been many changes in both product and the seafood business in the intervening years. About 70 percent of the material in this book is new, a tribute to the rapid pace of change throughout the industry.

The subject of this book is fresh seafood. "Fresh" is defined as product handled under refrigeration (mechanical or ice) from harvester to consumer. This excludes frozen product, canned product and other shelf-stable packaging. Frozen seafoods are covered in the companion volume, *The New Frozen Seafood Handbook*. Many products are, of course, handled in both refrigerated and frozen forms. There may be substantial differences, not just in how they are handled, but in how they are processed, graded and packed. Frozen seafoods are often treated and traded as commodities, with standard descriptions. Marketing and distributing fresh fish and shellfish, which has to be eaten within days of harvest, is necessarily more personal and direct.

The contest between refrigerated and frozen seafoods has continued for many years and shows no signs of resolving. Despite massive improvements in the quality of much frozen product, consumers and their retail and restaurant suppliers still tend to believe that "fresh" is better, perhaps simply because the word "fresh" is naturally appealing. Food writers and editors continue to promote "fresh" as superior to "frozen," without offering much in the way of supporting evidence. Given this prejudiced environment, it seems likely that consumers and distributors will continue to bear the high costs of handling refrigerated seafoods.

The general assumption is still that fresh seafoods are superior to frozen. Fresh seafoods, though, are harder to buy, more difficult to handle and less consistent than frozen fish and shellfish. Information remains

PREFACE

the foundation to success and profits. Understanding the seafoods that are available and the alternatives to them and knowing how to handle them and what problems to anticipate assist you in improving your bottom line. This book is designed to help intelligent seafood users survive and prosper.

The New Fresh Seafood Buyer's Guide

A manual for distributors, restaurants and retailers

How To Use This Book

This book is in encyclopedia format, with the entries arranged alphabetically. Headings include topics (such as handling) and products (such as scallops). For finfish product entries, we have favored the market name listed by the U.S. Food and Drug Administration (FDA) in its 1988 *Fish List*. Where there is basically one species or one dominant species on the market, the scientific name is given immediately after the title of the entry. Common names, which are also permitted by the FDA in interstate trade, are given next. Other names sometimes used, many of which are illegal, are listed where appropriate to help you locate descriptions from vernacular names. The *Fish List* is printed in the Appendix. It has been edited to correct a number of typographical errors in the original document. Information on family and location of the fish has also been added.

Shellfish names have not yet been codified by the FDA, but generally offer fewer confusions and complexities. Comments are included on equivalents for finfish market and common names where appropriate.

Compared with the first edition of *Fresh Seafood—The Commercial Buyer's Guide,* cross-references to other entries have largely been eliminated to save space. If you cannot find an entry, use the index to locate where the subject is covered. The index is full and comprehensive and it is there to be used and to be helpful. Where another heading has additional information, it is printed in CAPITALS.

1

A

ABALONE *Haliotis spp.* Also called ormer (mainly in the United Kingdom). Abalone are large molluscs with a single shell. The Latin name means sea ear, which indicates its shape: the shells are round or oval with a dome towards one end. Almost all of the limited U.S. production comes from California, which is also the largest market.

The following species are found on the west coast of the United States:

Haliotis assimilis Threaded abalone
Haliotis corrugata Pink abalone
Haliotis crackerodii Black abalone
Haliotis kamtschatkana Pinto or northern abalone
Haliotis rufescens Red abalone
Haliotis sorenseni White abalone
Haliotis wallalensis Flat abalone

Red abalone is the major commercial species, harvested in northern and southern California under strict regulation. It is now being farmed and supplies from aquaculture are increasing. Red abalone is the largest abalone species in the world, reaching 11 inches across, when it may weigh as much as eight pounds. Pinto abalone, which grows to five inches or so, is harvested in Washington, British Columbia and Alaska on a very small scale. In the past, northern production was canned, but now it is mainly sold fresh or frozen.

Abalone is highly prized in Europe and Asia. The European species, *Haliotis tuberculata,* has been over fished, although efforts to produce more from farming are under way. Japanese abalone, *Haliotis discus,* is similar to red abalone. Australia and New Zealand have a number of species. Commercially harvested abalone from these countries is sold in Asia.

Abalone are harvested by divers, who pick them individually from the kelp beds where they feed. There are strict controls on fishing, with limited seasons and bag limits. Stocks are small and carefully monitored.

Prospects for increasing supplies rest entirely on aquaculture, which is beginning to be successful in California, despite considerable difficulties in getting licenses and permits. In Japan, abalone farming is well established and successful. Hatchery techniques are well developed, so aquaculture prospects are limited only by the need to find suitable growing sites. Hawaii, which has no natural abalone resource, is being considered as a suitable production area.

The edible part of the shellfish is the foot, which provides a circular piece of meat, usually sliced horizontally across the grain to make two large disk-shaped steaks. The steaks must be tenderized before use. This is best done by pounding them lightly with a wooden mallet to break up the tough fibers. It is important to cook abalone for only a very short time or, like many other molluscs, it toughens. Fresh abalone is available as cleaned meats or as steaks, sometimes breaded and ready for the pan.

Californian farmed product is generally offered in the form of medallions, which are actually steaks produced from small abalone. These shellfish grow rather slowly, so harvesting them from farms when they are small makes economic sense. These small steaks have the same taste and texture as large ones and, if properly presented, provide better plate coverage.

Although California still provides most of the small U.S. market for abalone meat, the spread of sushi restaurants across the United States has increased demand nationwide.

Because abalone is one of the most expensive shellfish products, unscrupulous dealers are tempted to offer substitutions. Cuttlefish and giant squid have been presented as abalone steaks. The body meat, or mantle, is cut into circular pieces and then tenderized with a needle machine (normally used for softening cheap cuts of beef). Abalone showing a needle pattern is most unlikely to be the genuine article. Cuttlefish mantles are covered in a thin membrane. Small pieces of this remaining on the meat also indicate that the product is not abalone: abalone has no membrane.

A Chilean shellfish called a loco (probably *Concholepas spp.*) has in the past been imported and labeled as abalone. This shellfish is small and tough and quite unlike abalone in appearance, taste and texture.

ADDITIVES
Substances added to food for flavor, color, preservation, nutritional enhancement and many other good reasons. The FDA publishes a list of additives, including things like salt, designated as GRAS, which stands for Generally Recognized as Safe. Substances with GRAS status may be used (in certain ways only) without further ado. Other additives used in the United States are subjected to very long and rigorous testing before they are approved for use. Additives are an essential factor in our food supply: without anti-oxidants and other preservatives, a great deal of our food would spoil before it could be eaten.

Fresh seafoods may be dipped in phosphate solutions (see DIPS) to whiten them and reduce natural drip loss. Liquid smoke may be used for color and flavor. Sulfites, usually sodium bisulfite, are used to prevent melanosis of shrimp, though almost all of our shrimp is distributed frozen, not refrigerated. Sulfites are also permitted on clams, scallops, lobsters and dried cod. They may not be used on crabs or other seafoods recognized as sources of vitamin B_1.

The FDA has defined irradiation as an additive. Gases that may be used in packaging (as well as certain constituents of packaging materials) are also defined as additives. Partly for this reason (and partly because of fears that careless handling could cause the growth of botulinus organisms), American consumers do not have access to the gas-flushed packaging which has been remarkably successful in European countries in keeping chilled seafoods in excellent condition throughout the distribution chain.

All foods, including fresh seafood, that contain additives must be labeled to show the additive used. If you are selling scallops that have been treated with phosphates, the label or display ticket should state this. Foods which are not properly labeled are regarded as misbranded and are liable to seizure. For more information on additives, see the headings DIPS, GOOD MANUFACTURING PRACTICE and GRAS.

AIR FREIGHT Air shipment of fresh seafoods has become extremely important in the last decade. Fresh and even live seafoods are routinely shipped huge distances: live fish from New Zealand to Japan and the United States; fresh salmon from Alaska to Florida; sea urchin roe from Maine to Tokyo. The National Fisheries Institute (1525 Wilson Blvd., Arlington, VA 22209) and the Air Transport Association of America have produced guidelines for packing and handling seafoods shipped by air. Anyone shipping or receiving airfreighted product should read these guidelines, which cover packing, acceptability of product, acceptable refrigerants, labeling, packaging design, documentation and handling claims.

Live fish and shellfish require extra care. Success depends partly on the condition of the animal prior to shipping, partly on maintaining suitable conditions during transport and partly on careful handling to reduce stress.

Airlines seldom guarantee delivery at a specified time. If product gets delayed, it may spoil. Shippers should track their containers and ensure that they get prompt reports from customers on the condition of the product when it arrives.

ALEWIFE *Alosa pseudoharengus*. River herring is an alternative market name. Gaspareau, gaspergoo and gray, white or spring herring are not legally acceptable names. The alewife is an anadromous herring which spawns in the rivers of New England and the Canadian Maritimes in April and May. Alewives are about 10 inches long, slightly deeper in the body than a herring and silvery colored. There is a similar landlocked variety of the fish which lives in the Great Lakes, again swimming up the tributary streams to spawn. This fish is normally only about three inches long.

Both types of alewife have dark flesh and are dry tasting and very bony. Some ethnic groups eat them fresh or salted, but alewives are unlikely to find an expanded market among other consumers.

When utilized for human food, the fish is normally sold whole or dressed, though it can also be pickled. It could be used as a cheap pan fish in retail outlets, but its eating qualities are rather poor. Roes and milts provide better eating.

ALLIGATOR

American alligators were removed from the list of threatened species in 1986 after their numbers recovered to such an extent that the animals were becoming a nuisance in parts of Louisiana. Alligator farming is also flourishing in the Gulf Coast states. Although alligators are primarily hunted and farmed as a source of leather, the meat is finding increasing markets.

Alligator meat must be trimmed of all fat, tendons and soft tissue. This leaves small pieces of meat which are usually cubed and sold frozen. Small quantities of fresh alligator meat are available. Darker meat from the legs and belly is less expensive than the white meat from the tail.

AMBERJACK

Seriola dumerili. Also called greater amberjack. The fish is not a YELLOWTAIL and should not be labeled as such.

Amberjack is a jack. It is normally between 10 and 40 pounds, with some specimens reaching 100 pounds. The flesh is dark and oily, with a rather short shelf life. It is well flavored, especially when broiled, which is usually the best cooking method for oily fish, and it cooks up quite white. It is best if the fish is bled by immediately cutting off the tail as it is caught. This gives lighter meat and a more delicate flavor, as well as extending the shelf life. Because amberjack are not heavily fished, they may have some parasites, making them unsuitable for eating raw. The parasites can be located on a candling table and removed with a knife.

Most amberjack are caught on lines in the Gulf of Mexico, but the species ranges widely throughout the tropical and subtropical Atlantic, from the Mediterranean to west Africa in the east and from Chesapeake Bay (in summer) to Brazil in the west. There have been occasional reports of ciguatera in amberjack from the Caribbean, but this does not seem to be a problem elsewhere.

Amberjack is sometimes sold in California, where it is fraudulently substituted for yellowtail.

ANADROMOUS FISH

ANADROMOUS FISH Species which live in the sea but migrate to fresh water to spawn. Salmon are the best known anadromous fish as well as the most important commercially. Shad is another commercially significant example. Because anadromous species depend on very specific habitats for spawning and the survival of young fish, they are particularly vulnerable to the effects of pollution and industrial development.

ANAEROBIC BACTERIA Bacteria which live and multiply without oxygen. Canned and vacuum packed foods are vulnerable to contamination by anaerobic bacteria, which can grow inside a sealed can. The best known anaerobic bacterium is botulinus. This produces a toxin which causes the dangerous illness known as botulism poisoning. Botulinus prefers a low acid surface. Most seafood products are low acid foods. Salt deters the growth of botulinus, but the levels of salt required to make a product safe inside packaging which excludes oxygen are far too high for modern taste preferences.

There were two episodes of botulism from canned salmon in 1978 and 1982 which together virtually destroyed the canned salmon industry for a number of years. The problem should not be overemphasized: the number of cases of botulism from commercially produced seafoods has been extremely low. Because of the risk, the FDA does not permit modified atmosphere packaging to be used in the United States for retail packs of seafoods. Vacuum packs in the United States use film which allows the passage of oxygen.

Vacuum packed and retorted smoked seafoods are theoretically vulnerable to the growth of anaerobic bacteria, but good manufacturing practice and supervision has maintained the industry's clean record. Protection demands sanitary handling of seafoods to ensure that product being packaged does not become contaminated with bacteria and careful attention to storage temperatures throughout the distribution chain. Poor sanitation allows initial contamination by the bacteria; temperature abuse allows the bacteria to multiply. Clean packing conditions and good refrigeration by distributor and retailer can eliminate this risk altogether.

ANCHOVY Small, sardine-like fish, usually about seven to ten inches in length. Worldwide, there are over 100 species of anchovy and anchoveta, representing some of the largest volume fisheries. Most of the catch is made into fishmeal. Smaller quantities are canned or dried.

In the United States *Engraulis mordax*, sometimes called the northern anchovy, was once abundant off California. The resource appears to be recovering, but only small quantities are fished and almost all of the catch is used for bait. This species is generally about four inches long and works quite well as whitebait.

Anchovies imported from South America are occasionally available fresh in wholesale markets, but both these and the northern anchovy are unusual seafoods.

Anchovies have very soft flesh and, if available, are offered whole. Both features contribute to a very short shelf life. Handle them with care, making sure that ice does not puncture the delicate flesh. Use them quickly, before they decompose and develop rancidity.

ANTIBIOTICS Antibiotics such as aureomycin and tetracycline are used in some countries to help preserve fresh fish. The antibiotics are added to ice and work by killing the spoilage bacteria on the surface of the fish. Because small quantities of antibiotics remaining on the fish can build resistance to the drugs, the practice has always been banned in the United States. The FDA monitors imported foods and will reject any that show traces of antibiotics.

The use of antibiotic drugs in fish farming is similarly proscribed. Many medications are used to treat young fish, but the basic rule is that there must be no trace left by the time the fish are harvested. The FDA approves only three drugs for use in farmed salmon and there are small tolerance levels for two of them (meaning that traces at or below these levels in harvested fish are permitted):

Oxytetracycline (OTC)	0.1 part per million
Sulfamerazine	zero tolerance
Romet-30	0.1 part per million

Salmon farmed in Norway, Chile and other countries has to meet similar or stricter standards.

A.O.A.C The Association of Official Analytical Chemists. This is the body which approves scientific testing methods for standard use. Tests done in a manner approved by the AOAC are generally recognized as conclusive evidence in both federal and state courts. The methods are tested by arranging for a number of different laboratories to carry out identical tests on identical samples. If they all get the same results, then the test is sufficiently reliable to be published as a standard, approved method. If you buy testing services of any type on your seafoods, ensure that the laboratory uses only AOAC approved tests if these are available.

AQUACULTURE The latest definition from the FAO is "The farming of aquatic organisms...Farming implies some form of intervention in the rearing process to enhance production... [and] individual or corporate ownership of the stock being cultivated." Aquaculture is the fastest growing sector of the fishing industry and the major hope for continued expansion of supplies to meet growing markets.

Fish and shellfish have been farmed for thousands of years. Oyster farming can be traced back over 2,000 years to Roman enterprise. Carp and milkfish have been Asian staples for perhaps as long. Aquaculture, like so many industries, has now moved technologically far ahead of these early examples. One of the most important keys to successful aquaculture (or to successful land-based animal husbandry) is a reliable supply of young animals to raise. Development of reliable hatchery techniques for many species has made modern aquaculture possible.

Aquaculture products are important in a number of United States markets. Shrimp aquaculture in 1989 produced 24 percent of the world's commercial shrimp supply of 2.2 million metric tons, according to *World Shrimp Farming*. China produces about 30 percent of farm raised shrimp, followed by Indonesia, Thailand, the Philippines and Ecuador, which is the leading producer in the western hemisphere and was the early pioneer of shrimp aquaculture techniques. In the same year, over a quarter of all salmon came from fish farms, representing startling growth from an industry that barely existed at the start of the decade.

Most oysters sold in the United States are farmed (at least on the rather broad FAO definition). Catfish is almost entirely farmed, the aquaculturists having swept most of the market from the generally inferior wild, imported product. Farmed fish similarly dominates the trout market, not just in the United States but worldwide.

In Japan, where seafoods are far more important to the overall diet than they are in the United States, scallops, yellowtail and hundreds of other species of shellfish and finfish are farmed successfully.

The United States lags far behind other countries in aquaculture, largely because of institutional constraints on development and investment. Most fish farming requires sea water and there are many competing demands on scarce shoreline which limit the opportunities for farmers. Permits for fish farming are difficult to obtain in all the western states and in most of the eastern ones. Alaska in 1990 permanently banned the farming of marine finfish in the state, despite the enormous potential for salmon aquaculture. The major reason for the ban was political pressure from fishermen who fear the loss of traditional markets to less expensive and more consistent aquaculture product.

Aquaculturists are now experimenting with many different species. Bay scallops, which have not to date been profitably grown in their native United States, have been introduced to China and have already formed the basis of a new industry. Turbot (the genuine, expensive European flatfish, not greenland turbot) is being farmed in France and Spain. Even the once-inexpensive cod is being grown in Norway.

Ranching, which is a variation of fish farming where the fish are left to their own devices to feed in the ocean, is being applied with, so far, only moderate success to salmonids, including specially bred steelhead. If successful, ranched salmon could feed freely and for free in the open ocean and then return for harvest.

The implications of aquaculture for the fresh seafood user are profound. Firstly, aquaculture promises, and frequently delivers, more uniform product. When shrimp ponds and salmon pens are harvested, the contents all tend to be about the same size. Aquaculture also offers some protection from seasonal fluctuations. Although natural cycles can never be totally eclipsed, harvest seasons for many species can be extended and the possibilities of growing a northern species in the southern hemisphere also extends the possible harvest period. Salmon, for example, is harvested from Chilean farms in the northern winter when there are few fish harvestable from natural, wild stocks and not too many available from northern hemisphere salmon farmers either. Fish farmers are working with geneticists to breed fish with more desirable traits, such as faster growth (which reduces costs). Farmed steelhead have been bred with red flesh and deep bodies to suit particular market requirements for salmon. Oyster growers have developed strains with deeper cups, better meat yield and even an inability to develop sexually, so that they can be eaten all year round.

ATKA MACKEREL

Pleurogrammus azonus and *Pleurogrammus monopterygius*. This is not a mackerel, but a greenling and a relative of the lingcod. It is one of the four most abundant fish of the North Pacific, is well utilized in Japan, and generally regarded as a trash fish in the United States, Soviet, Japanese and South Korean vessels take about 200 million pounds a year of *P. azonus* from the northern Pacific and the Bering Sea. Salted Atka mackerel was sold in the last century to gold miners in Alaska and California, but there is little or no current market interest in Atka mackerel in America. The flesh is rather oily and dark colored but, according to NMFS researchers, flakes well and has a good flavor. It reportedly makes good smoked and canned products. The fish is only about 14 inches long, which means that although it can be filleted, the fillets are rather small. It is a brightly colored, very attractive fish.

B

BARNACLES *Pollicipes spp.* Normally called goose barnacles or goose-necked barnacles. These creatures are crustaceans, although they look much more like molluscs. Goose barnacles have a long neck or stalk, the peduncle, which is boiled, skinned and eaten. *Pollicipes pollicipes* is the European goose barnacle, especially popular in Spain and Portugal. *Pollicipes polymerus* is found from Sakhalin through the Aleutians and along the Pacific coast of North America as far south as Baja California. It is much larger than the European barnacle, reaching over four inches, but it is similar in taste and texture and quantities are exported from British Columbia and the United States when they can be harvested. There is also a tropical species found from Mexico to Peru and another from New Zealand. None of these species is abundant.

Barnacles grow on rocks drenched by ocean spray, which makes then difficult to harvest, because they are hard for fishermen to reach. They adhere strongly to the rocks and have to be cut off one by one. It appears that once a cluster of barnacles is damaged by harvesting some of them, the whole colony takes many years to recover. Sometimes mussels invade the area and grow so fast that there is no room for the barnacles to reestablish themselves. Consequently, harvesting goose barnacles is not only difficult, but is strictly controlled in order to protect the fragile resource.

Barnacles are also easily damaged in harvesting. The peduncle houses most of the important organs and is easily crushed. Barnacles have to be alive when cooked. Storage and transport require carefully controlled, damp and cool conditions (about 45°F). Barnacles last much longer if the whole cluster, together with the rock they are on, can be harvested so that they are not damaged by being removed from their positions.

BARRACUDA Pacific and Atlantic species have to be treated quite differently.

Pacific barracuda. *Sphyraena argentea.* This fish is also called snake, scooter and California barracuda, but none of these names are recognized for interstate trade.

Caught in spring and summer off California and available from Mexico for most of the year, this is a popular food and game fish regionally. It is usually sold fresh or occasionally smoked or dried. The flesh is firm and suitable for broiling and barbecues. Most fish caught are around five to eight pounds. Headless, dressed fish are the usual market form. Barracuda has a poor shelf life and should be used as quickly as possible. It has a better shelf life and flavor if it is bled when caught.

Availability varies considerably from year to year, probably because the fish migrate according to ocean temperatures.

Great barracuda. *Sphyraena barracuda.* Commonly but incorrectly known as the Atlantic barracuda, this fish can cause ciguatera poisoning (for a full discussion, see MARINE BIOTOXINS, below). Fish caught around the Caribbean islands, especially larger ones, may contain the toxin. Florida fish seem to be less affected, but should also be avoided.

Since there is no way to be sure whether the fish is poisonous or safe, the only sensible course is to avoid handling Atlantic barracuda at all.

European barracuda. *Sphyraena sphyraena* is a smaller fish which is sometimes caught off the Atlantic coast.

BASS The name refers to many different and unrelated fishes, from fresh and seawater. Fish described as bass commercially include grouper, black sea bass, jewfish, STRIPED BASS, WHITE BASS and drum. The channel bass is also called redfish and is described under drum. See the various relevant entries. The true bass is a western European game fish prized particularly by surf fishermen.

WHITE SEABASS: *Cynoscion nobilis.* The FDA prefers the market name of seatrout. The fish is a drum, not a bass. It is a large fish, often 15 to 20 pounds, caught commercially in California and Mexico and sold dressed or as fillets. It is mainly available during the summer months.

BLACK SEA BASS: *Centropristis striata* is a reef fish, sold whole or filleted. It is sometimes described as rock bass or as blackfish. Neither of these names is acceptable to the FDA.

Mainly caught and landed between Long Island and North Carolina, the black sea bass generally runs about one to three pounds. It has poisonous spines on its back and must be handled with care. Smaller fish are popular in Chinese cuisine for serving dressed and deep fried, or with sweet and sour sauce. Black sea bass has firm, white flesh and is versatile to cook and use. The males develop a hump-like shape behind the head and are sometimes called humpbacks.

Supplies are available from Florida through the winter and from the middle Atlantic area between September and March from offshore fishing and between May and November from traps.

Parasitic worms are a substantial problem with black sea bass and account for its lack of popularity. The fish should be bled as soon as it is caught. The dorsal fin has sharp spines; it should be completely removed before the fish is offered for sale commercially.

BELLY BURN   The condition of a gutted fish where the flesh of the walls of the belly cavity is pulling away from the bones of the belly cage. Fish guts contain enzymes which rapidly break down flesh adjacent to the intestines if the carcass is left for too long with the guts in. Belly burn is a sign that fish has not been properly and quickly gutted and chilled. There are other possible causes such as rough handling (picking the fish up by the belly walls or standing on it, for example). The presence of belly burn definitely indicates that the fish is less than the best quality.

BILLFISH   This term taxonomically refers to spearfish, sailfish and marlins, but most people include swordfish as well. Apart from swordfish, most billfish are basically recreational species and as fishing pressure increases are more likely to be reserved for sport fishermen.

BISQUE   Thick soup made traditionally from lobster, shrimp or crabs. Bisques are also made from scallops, various fish and seafood in general. Traditional bisques are thickened with butter and cream.

Good bisque is not difficult to make from small quantities of suitable materials. It is also possible to buy prepared bases, which require water or other liquid to be added, as well as ready-to-serve bisques.

BLEEDING   Blood remaining in the muscle (fillets) of fish is often considered a defect. Blood clots and bruises reduce the value of white fish and severely affect the usability and yield of salmon that is to be smoked. Bleeding fish soon after it is captured avoids these problems and often increases the value of the fish. With many species, bleeding also improves the shelf life of the product, especially if it is distributed fresh. Blood naturally contains a great deal of oxygen. If the blood is removed from the fish, oxidation (rancidity) is slower, so the fish remains palatable longer. There are few if any circumstances when it would be preferable to retain the blood in the fish.

Fish blood remains fluid at chill temperatures for about 30 minutes after the fish is captured, which is not very long. During this period, if the fish is gutted, some of the blood will naturally flow out of the gut cavity. Other fish may be bled by cutting off the tail immediately after it is landed on deck. Bleeding sharks in this way is particularly important as the heart continues to pump blood out of the carcass. Sharks carry uric acid as well as oxygen in the bloodstream. If this is not removed the fish will develop an ammonia taint, which greatly reduces its commercial value. Farmed salmon are sometimes stunned or asphyxiated and then bled by cutting the gill notch. This produces particularly good quality meat. Salmon is an oily fish which can turn rancid rather quickly unless it is bled.

In summary, bleeding the catch should be a normal requirement of every finfish harvester, because bleeding improves the shelf life and appearance of the product.

BLOATER
1. Cold smoked, whole, fat herring. See SMOKED FISH. 2. Name sometimes used for cisco. 3. Name sometimes used for smoked chub.

BLUEFISH
Pomatomus saltatrix. Names such as tailer and chopper are not approved. Small bluefish are often called snappers. Sablefish may be called bluefish, though this use is rare. Boston bluefish is not bluefish at all, but pollock.

Bluefish are a large, oily relative of the mackerel, caught in large numbers by sport and commercial fishermen as they migrate up and down the Atlantic coast. Commercially available fish are generally between two and eight pounds, while fish as large as 20 pounds are not uncommon.

The supply of bluefish is substantial, but fears that the resource may be stressed prompted the initiation of a bluefish management plan covering the fishery from Maine to Florida. Under the plan, 80 percent of the catch is reserved for sport fishermen who generally sell most of their catch to wholesalers and dealers, so it ends up in the commercial area. Surprisingly often, especially in New England, dedicated bluefish fishermen will not eat the fish themselves, regarding it as trash.

There are substantial bluefish resources in the Gulf of Mexico, which are hardly fished because of lack of local markets.

Bluefish migrate northward in early summer and back south in September, though the seasons vary according to the weather pattern and water temperatures for the year. Supplies are available from the Gulf side of Florida between November and March and from Atlantic Florida all year round. Peak production in the middle Atlantic states is usually August and September, but the fishery can be highly local and dealers in one area will insist there are no blues while wholesalers only a few miles away may have more than they can handle. Intelligent buying of

BONELESS FILLETS

fresh bluefish therefore requires keeping track of the movements of the fish and dealing with the current areas of plenty.

It is essential that bluefish be handled swiftly and properly. The fish must be immediately gutted and bled as it is caught and then be well iced. The flesh turns rancid very quickly—everything possible must be done to delay this process. Unfortunately, because so much is caught by sportsmen who are interested in catching the fish rather than in its value to the seafood trade, quality of landed bluefish is erratic. When the fish are running, they come in large numbers and there is no time to gut and ice them unless someone is willing to stop fishing. Consequently, it is particularly important that dealers examine blues they buy from the fishermen and that users get to know which dealers take that extra and necessary care.

Bluefish are not much sold inland: the coastal areas where they are available straight from the water make up most of the market. Since the meat deteriorates so quickly, this is understandable. Bluefish can be frozen but must be very fresh for successful results. Frozen bluefish has a short storage life, again because it turns rancid easily.

Bluefish flesh is dark, but when cooked is fairly light colored. The flake is large, but the meat is rather soft. Flavor is strong, often described as fishy. Larger fish have stronger flavor than smaller ones. Smoked bluefish fillets and products such as pate made from it are gaining increasing acceptance. Bluefish is excellent smoked and comparatively inexpensive. It deserves to be better known.

Nearly all bluefish is sold fresh, either dressed or filleted. Fillets turn rancid even faster than the dressed fish, so blues are normally filleted at a late stage in the distribution chain. Bluefish is quite cheap and if it has been properly handled offers excellent value for money.

BONELESS FILLETS Many seafood users assume that fillets are pieces of fish without bones. This is unfortunately not correct. A fillet is a side muscle from a fish (see FILLET for a more complete definition) and most species of fish have a row of pinbones which run horizontally along the center of the fillet from the nape (head) end to about one third of the way down. These pinbones can be cut out, but the yield is significantly reduced because fish has to be cut away with the bones.

Boneless fillets are defined by the National Marine Fisheries Service (NMFS) in Title 50 of the Code of Federal Regulations (part 263.104). The standard allows for a few instances of bones.

The pinbones of most smaller fish soften when cooked and are practically undetectable. Flounder, sole and other flatfish do not have pinbones, which may be an important reason for their popularity. Consumers seem to be terrified of bones in fish and dealers attempt to supply boneless products.

If you want boneless fillets, you must so specify or cut out the bones

14

yourself. Fresh fillets such as cod and haddock are rarely if ever shipped boneless, although boneless products are made for frozen sale. Fish such as snapper are filleted in a different shape and so can be shipped in boneless form.

BONELESS COD
A confusing term which refers to salted cod of a high grade, packed without skin and bones. Semi-boneless means that some of the smaller bones remain.

BONITO
Sarda sarda, Atlantic bonito and *Sarda chiliensis*, Pacific bonito. These fish are sometimes incorrectly called frigate mackerel. Bonito and frigate mackerel are very similar.

Bonitos are tuna-like fish reaching about seven pounds in the Pacific and 12 pounds in the Atlantic. The two species are identical for commercial purposes. Like all oily fish, bonito turns stale and then rancid swiftly. Bleeding, fast handling and plenty of ice are essential.

Although bonito is frequently canned, it must not be labeled as tuna. The meat is fairly dark and oily, more like mackerel than tuna. There is a substantial market along the east coast and in Puerto Rico for fresh bonito, usually sold headless dressed or steaked. The steaks broil well, have good flavor and texture and the meat can easily be removed from the large bones.

BOSTON BLUEFISH
A name used regionally for Atlantic pollock, which is not in any way like bluefish. Since bluefish is normally cheaper than pollock, the purpose of the misnomer is unclear.

BREAM
This name is applied to a great variety of fish. Ocean perch (redfish) is sometimes called sea bream. Scup is also occasionally called sea bream. It is totally different from the redfish. The European (fresh water) bream, or bronze bream, is not known in the United States but many other freshwater species, mainly sunfishes, are called bream locally, especially in the south. Crappies, sunfishes, rock bass, black bass and pumpkinseed are some of the fish that may be called bream. While few of these are used commercially, some may be locally important. The only way to define your bream is to have a look at the fish itself.

BUCKLING
Lightly cured, hot smoked, dressed herring. See SMOKED FISH.

BUFFALO
Also called sucker, but the name is not an approved one. There are numerous species, mostly members of the *Ictiobus* group. Bigmouth buffalo (*I. cyprinellus*), smallmouth buffalo (*I. bubalus*) and

black buffalo (*I. niger*) are the most important. *Catastomus spp.* and *Moxostoma spp.* are also sold as buffalo or sucker.

Buffalos are fish similar to catfish in taste. Generally marketed at three to five pounds, they may grow as large as 45 pounds. They are one of the most important freshwater commercial fish, especially in the midwest and the south. Large quantities are grown in ponds.

The flesh is white and sweet, and very lean. Buffalos have a lot of bones, but consumers nevertheless continue to like the fish. Most fish are sold dressed.

Suckers from Canada are frequently labeled and marketed as mullet, which they do not resemble in any way.

BURBOT   *Lota lota.* A fresh water member of the cod family, found throughout the colder regions of North America. Little utilized commercially, the burbot is usually about one to two pounds, except in Alaska where some fish have been found over 50 pounds. It is sometimes wrongly called freshwater cod.

BUSHEL   A volume measure equal to 32 quarts, or eight gallons. Used for molluscan shellfish such as clams, oysters and mussels. Also used for blue crabs. The shape of a container can affect the quantity of product that it will hold. Consequently, a number of states with important shellfish harvests mandate the shape and size of standard bushel containers.

It is not possible to relate shellfish weights to volume measures. A bushel of mussels, for example, may weigh as little as 45 pounds or as much as 60 pounds. Relative shell and meat weights vary with the season, the place of origin and other factors.

BUTTERFISH   *Peprilus triacanthus.* Also wrongly called harvestfish, dollar fish and silver dollar. This is a small oily fish, usually around three to five ounces but sometimes growing to 20 ounces, abundant in spring and late fall off the northeast and middle Atlantic regions. It is shaped rather like a pompano (and is sometimes called pomfret in Europe). The Gulf butterfish, *Peprilus burti,* is a very similar fish. In recent years it has been more abundant than the Atlantic species.

A related Pacific species, *Peprilus simillimus,* is permitted to be called Pacific pompano. This species is sold in small quantities throughout the year in California. Sablefish (*Anoplopoma fimbria*) are sometimes called butterfish in California, although there is absolutely no similarity between sablefish and butterfish. The use of the name for sablefish is illegal in California and federally.

Butterfish are sold whole, fresh or frozen. Fresh fish may also be headed and dressed, possibly tray packed. They are also smoked. Butterfish are good (if small) panfish. The flesh is fat but turns white when cooked and has excellent flavor. The numerous small bones have prevented the species finding wider markets among American consumers. Some seasons, Japanese buyers take huge quantities from the Atlantic states, especially from Rhode Island, where the greater part of the U.S. catch is landed. They take none at all in other years so although there is generally plenty of butterfish in the sea, supplies reaching domestic markets are very erratic and impossible to forecast.

The Japanese market prefers the fall season fish, which are free of roe and have the highest fat content. The domestic market is rather limited, but smoked butterfish appears to be gaining favor slowly. Dressed fish are cold smoked and then hot smoked, resulting in a product with bones softened sufficiently as to make them barely noticeable.

Fresh butterfish, shipped whole, has a shelf life of only a few days. Fish from traps is often a better buy because it has been treated more gently than the bulk of the catch, which is landed by trawls. As the fish ages, the silvery skin turns gray and dull.

C

CALAMARE Italian word for SQUID. Spanish is calamar and French is calmar, though there are other words in each language for different types of squid. The name calamare has become useful in marketing squid in the United States because it obscures the fact that the product is squid—a word which seems to frighten many consumers, who may be persuaded to try it if they do not know what it is.

CAPELIN *Mallotus villosus.* A smelt from northern polar waters, found in both the Atlantic and Pacific oceans. The Newfoundland cod industry was based on capelin: the cod were caught as they followed the huge runs of spawning capelin onto the beaches.

Capelin are very oily. They can be used in much the same way as smelts, although they are not particularly palatable. Use of fresh capelin is limited to areas close to where the fish is caught. Capelin roe is an important product in Japan; Canadian packers export female capelin frozen for this market.

CARP *Cyprinus carpio.* This fresh water fish has been farmed for centuries in Asia and Europe. It was introduced, perhaps unwisely, into the United States about a century ago and spread through much of the country, occupying many lakes and streams. Substantial quantities are available from aquaculture as well as from regular wild sources. Although there are various varieties, such as mirror carp and leather carp, these are all strains of the same species and flesh quality relates more to where the fish is grown than to the precise breed of the carp. The grass carp or white amur, *Ctenopharyngodon idella,* is becoming widely available from catfish farmers, who can add them to their ponds and grow them with no additional feed.

Carp grow as large as 50 pounds, though about five pounds is more usual for commercial purposes. Markets are chiefly among people of central European and Jewish origin. Because kosher rules require positive identification of the fish, it is generally shipped with the head on to ensure proper classification.

Carp have to be skinned before use. The flesh is light tan (in female fish) to reddish (in males). Both sexes have a band of dark fat along each side which must be discarded: it is tough and rather unpleasant. The remaining light meat, however, is appetizing and versatile.

CARPET SHELL
A name used for various types of hard shell clam. The term is not normally used commercially in the United States, but is important in international statistics and trade.

CATADROMOUS
Catadromous fish spawn in sea water but live most of their lives in fresh water. Eels and barramundi are catadromous. There are few other commercial examples. Fish which follow the opposite pattern, living in sea water and spawning in fresh water, are called anadromous. There are many more of these.

CATFISH, FRESHWATER
Ictalurus punctatus. Also called channel catfish. Bullheads (*Ictalurus spp.*) are very similar, effectively interchangeable in use. For spoonbill catfish, see PADDLEFISH. Catfish from the Brazilian Amazon, *Brachyplatysoma vaillanti*, is readily available frozen. The fresh market is dominated by farmed catfish, usually distinguished by marketers calling it either channel catfish or farm-raised catfish.

There are many wild catfish species in the United States and worldwide, ranging as large as 300 pounds (in the Mekong as well as the Danube). Although farmers have so far concentrated largely on the one species, there are others which might be farmed in the future.

Aquaculture's greatest success so far in the United States is the development of the catfish industry. It has grown fast for a number of years and has attracted several large companies with the ability to spend money on effective marketing. Catfish now sells throughout the United States, thanks to the industry's concentrated and effective marketing efforts.

The domestic catfish market has seen an explosive growth in supplies in recent years, thanks to the development of techniques for farming them. Aquaculture methods of growing catfish ensure a consistent product which can, if required, be grown to precise specifications of flavor and size. Catfish farming is able to produce a fish with the minimum of 'fishy' odor. Fat content of the farmed product is substantially lower than that of wild catfish. This ensures the greater consumer appeal of whiter flesh and lack of fish flavor.

Comparatively small amounts of wild catfish are caught, mainly in the South. Most of this is recreational catch and is used by the fishermen themselves. Commercial supplies could be increased from areas such as Lake Okeechobee in Florida.

Fresh domestic catfish competes with frozen catfish imported from Brazil and with small quantities of farmed catfish from Mexico. In the main, if you want fresh rather than frozen product, domestic farm raised catfish is what you will get.

Catfish is regarded by many government entities as part of the agricultural sector rather than as a fishery. This has helped the industry raise development money. If you need to track production and sales statistics, contact the U.S. Department of Agriculture rather than the National Marine Fisheries Service.

Although fish farming makes it possible to supply fresh catfish throughout the year, peak production is August through October.

Farmed catfish are generally delivered alive in tank trucks to the processing plants, which are automated and handle this single product. The fish are killed and immediately processed: there is no need to hold the dead fish on ice for days while the boat continues to fish. Heads, skin and guts are removed and about half of production is sold fresh, either in this form or filleted. Many plants produce retail packs for supermarket sale, controlling the entire process from slaughter to pack in a single, continuous operation. Increasing quantities of catfish are being used in ready to cook and other added value preparations. Many of these are available fresh as well as frozen.

Headless and dressed, skinless fish comprise the major product form. The preferred size is around 10 ounces and most product is between 8 and 12 ounces. They are sold tray packed for retail sale and in vacuum packs of 30 pounds, which have an extended shelf life of about three weeks. Skinless fillets are also produced but are mainly frozen, as are steaks and a wide range of other products. Because of the standardization of the catfish business and the emphasis on brands, there are few problems with the product, compared with many other fish.

Farmed catfish are usually marketed at about one pound, live weight. The species will grow, in the right conditions, to 50 pounds or more, but bigger fish present a marketing problem, as well as additional production cost. Because so much catfish is farmed, supply adjusts quite rapidly to demand and year round availability is virtually ensured.

Wild catfish is a less certain product, more varied in taste and texture than the farmed product, which is now engineered for particular taste preferences (especially by removing any trace of fishy flavor). It may be offered dressed, but it is essential also to remove the skin before cooking the fish. Catfish have a band of fat running along the lateral line on each side of the body under the skin. If this is removed from wild catfish, the product tastes milder and is generally preferred. Because farmed catfish have less fat than wild ones, this defatting of farm raised catfish is not necessary.

Despite the ability to program production, catfish prices remain quite volatile. Perhaps this is to be expected in any farming operation: the pig cycle of classical economics is the illustration. Catfish farmers who find that markets are dull also find it difficult to store their product. If they keep it alive, it not only consumes food but also continues to grow and larger catfish are not as readily marketable as smaller fish. It is therefore preferable to process and freeze the fish, which again adversely affects the market for all catfish. Despite such problems, the catfish industry continues to grow and is in a better position than almost any other segment of the seafood business to compete with poultry and meat for consumer markets.

Catfish is generally fried. Properly cooked (not overbreaded and not overcooked), it is most palatable and excellent value for money.

CATFISH, OCEAN

Anarhichas spp. Also may be called wolffish. Large (up to 40 pounds), deep water northern Atlantic fish which feeds on shellfish and has particularly sweet flesh, similar to that of genuine Dover sole. Small quantities are landed in New England and Canada, but most of it is frozen for export to European countries, which regard it highly. Some frozen ocean catfish is available, but the item is seldom seen fresh in distribution. Cusk has similar flesh and is sometimes sold as ocean catfish.

CATFISH, SEA

Arius felis. Also called hardhead. This fish is taken by shrimp trawlers in the Gulf of Mexico as an incidental catch and it is found as far north as Cape Cod. Fish generally between 10 and 20 ounces are sold either whole, or headless and dressed. Although it is a good, mild tasting pan fish, it is little regarded and sales are restricted to places near to the landing areas. Poisonous spines on the fins make sea catfish an unpleasant fish to handle and it is too small for efficient machine processing.

There is probably a large resource available and sea catfish may offer opportunities for development.

CAVIAR

Eggs of sturgeon, salted to preserve them. In some forms, probably the most expensive seafood product. The USSR and Iran produce most of the world's supplies, from sturgeon caught in the Caspian Sea. Small quantities are produced domestically, from various species of sturgeon. Farmed sturgeon may supply caviar soon. Domestic caviar is normally cheaper than imported and is often better, because it is fresher.

The FDA requires that only sturgeon eggs be labeled caviar and that eggs from any other fish must have the name of that fish included on the label. Consequently, names such as salmon caviar or lumpfish caviar are often seen. These should not be mistaken for caviar. The word on its own is reserved for sturgeon eggs, not those from any other fish.

Most caviar is shipped chilled, though technological developments now make it possible to freeze caviar without bursting the eggs. The amount of salt used varies. Malasol caviar is least salted and may strictly be called fresh. All other types are preserved by the addition of between three percent and six percent salt. Caviar should be kept refrigerated between 26°F and 32°F (the salt content makes its freezing point slightly lower than this).

Characteristics of quality in caviar include whole eggs covered with their fat (containers should be turned during storage to ensure that the fat does not rise to the top) and absolutely no fishy odor. Size and color of eggs are not, strictly speaking, quality characteristics, though they may be important in terms of market preferences. Broken eggs are made into pressed caviar which has less eye appeal but can be equally good to eat. The USSR quality grades for caviar are Extra (the highest quality) Grade 1, Grade 2 and Pasteurized.

Types of Caviar

Malasol indicates "fresh" caviar, lightly salted. It can apply to any type of caviar and is not a type itself.

Beluga is made from the beluga sturgeon. It has large gray to black eggs and is generally the most expensive type.

Sevruga is made from the sevruga sturgeon, which gives much smaller eggs, similarly gray to black.

Osietr (or osetra) sturgeon produces a brown or gold caviar which is less often seen in the United States Note that "golden caviar" once came from the sterlet, now virtually extinct, and this form used to be the most expensive and rarest caviar made.

Pressed caviar is made from broken eggs and often has all the flavor of the whole egg product.

Caviar Substitutes

Red caviar is not caviar at all, but is made from salmon eggs and should be labeled as salmon. This is a truly excellent item. Chum salmon makes the best product, followed by coho and pink. King salmon roe, although it has large eggs, is not as good for this purpose.

Paddlefish roe is apparently being used to produce a caviar substitute domestically.

Lumpfish caviar is made from the roes of lumpfish (also called henfish) and is dyed black. It is totally unlike caviar and barely edible.

CAVITATION When ice melts away from the containers and fish which it surrounds, leaving an air space, the occurrence is described as cavitation—the formation of a cavity in the ice. It is important to prevent this because air is not a good insulator and the ice has to be in contact with the product if it is to keep the product properly chilled. Ice should be changed frequently to ensure that new ice is put into physical contact with the product.

CEPHALOPOD Octopus, cuttlefish and squid are the commercially utilized cephalopods, which are molluscs with heads and arms. In these animals, the shell of the mollusc has evolved into a small, internal cartilage-like structure, called the pen in squid and the bone in cuttlefish.

CHAR *Salvelinus alpinus*. Usually called by its common name of Arctic char. The alternative spelling charr is equally acceptable. We follow the practice of the American Fisheries Society, which prefers the shorter spelling.

Char, a salmonid, is closely related to trout and salmon. They are found throughout the world's arctic regions, including Siberia, northern Europe, Canada and Alaska. They are, like salmon, anadromous, but grow very slowly. They return to their rivers each fall and to the sea each spring.

Char come mostly from small, isolated populations, so they vary considerably in their basic characteristics. Flesh ranges from red through pink to white. The size of mature fish can be from under a pound in some landlocked populations to the normal two to eight pounds. The largest char recorded was 35 pounds (from Siberia). Although the fish is mainly important to local populations, there is usually a small commercial fishery off Labrador in the late summer. Dressed fish, fresh or frozen, is available from this fishery.

Aquaculture of char is showing early signs of success in Iceland and Norway. Small quantities have been sold in the United States. Canadian producers are also planning increased production of both large and pan sized fish (which can be shipped and displayed in live tanks like trout).

Handle and use the fish just like salmon. Many experts consider the taste and texture of a large, red-fleshed char to be far superior to salmon. If aquaculturists are successful, more people may have the opportunity to test this view for themselves.

The Dolly Varden, *Salvelinus malma*, is a very similar fish, now reserved for recreational fishermen. It is caught along the Pacific coast north of central California and on the Asian side as far south as the Korean peninsula.

CHOWDER Thick soups based originally on clams. Fish and corn chowders are also popular. New England clam chowder is made with milk or cream. Manhattan chowder is made with tomato juice. In Rhode Island, chowder may be a clear broth, a preparation which might gain more attention in the future from those concerned with the health effects of cream. There are thousands of variations and recipes.

Minced clams and clam juice may be used for making chowders, to avoid the work of steaming and opening clams. Although most chowder is now made from quahogs, mahogany or surf clams, any clam can be used. Ready-made chowders and chowder base are widely available, fresh and frozen. The general rule is to find a source or recipe you and your customers like, and stay with it. There can be enormous differences between chowders, which is not to pass judgment whether one is better than another. Most of the time, consumers prefer consistency.

The word is also a designation for a large quahog (hard clam) suitable for use in making chowders.

CIGUATERA A type of food poisoning caused by toxins in the flesh of certain fish in certain locations. It is not related in any way to the freshness or staleness of the fish and there is no way to tell from the appearance or smell of the fish whether it may contain the toxin. It occurs chiefly in parts of the West Indies and South Pacific, including Hawaii, where most of the U.S. cases are reported. Fish which are perfectly safe to eat elsewhere may be poisonous in these areas, but certainly not always. Among the fish which may be affected are the Atlantic barracuda and amberjack and some grouper and snapper species.

Symptoms of ciguatera poisoning include upset stomach, nausea, vomiting and numbness around the mouth and lips. Severe cases can cause convulsions. Ciguatera is rarely fatal, but it is nasty and can last for weeks.

The ciguatoxins are produced by dinoflagellate algae consumed by fish low on the food chain. These fish are in turn eaten by larger fish which are used as human food. The toxins are transmitted up the chain of predation, but do not appear to harm fish. The toxins are not destroyed by freezing or by cooking. Fish species from areas of ciguatera occurrence and which are known to cause ciguatera should be avoided.

CISCO *Coregonus spp.* These fish are also called tullibee and lake herring, as well as chub. All of these names are permitted in interstate trade. Cisco is caught in freshwater lakes in the midwest and New England and (mainly) in Canada. Dressed fish, usually weighing about one to one and a half pounds, are available from Canadian packers, though most of their production is frozen or smoked. By the time it reaches the consumer, chub is invariably smoked because smoking softens the many small bones and makes the fish palatable. It is sold as smoked chub or bloater.

CLAMS
Bivalve molluscs (shellfish with two shells) of many different types. The major commercial clams are discussed individually below. Fresh (wild) clams are a major seafood commodity. Supplies tend to be declining because of the pressure of fishing and because of increasing pollution and alternate uses of the tidal and shallow water areas necessary for most species' survival. Most conservation rules aim at making it as difficult as possible to collect clams, by limiting the amount that can be taken in a day and by restricting the fisherman to the simplest of tools and gear. Although this possibly serves a social purpose by protecting traditional clamming occupations in rural areas, it certainly does nothing to encourage more efficient development of the resources, which would include commercial planting and harvesting—in aquaculture–type operations—on a large scale.

Clam aquaculture technology is well developed. Although institutional constraints have restricted the development of the industry in the United States, there are now reasonable prospects that cultured clams could make a significant contribution to supplies.

Clams must be alive up to the moment that they are shucked or cooked. Dead clams will rapidly grow huge bacteria colonies which can be highly dangerous. Softshell clams show that they are alive by pulling the siphon as far as possible into the shell when they are touched. Touching the siphon usually causes the clam to retract it immediately. A dead softshell clam has the siphon hanging out and limp. Hard shell clams will also close up tightly when disturbed. Dead clams' muscles, which hold the shells together, relax and allow the shells to be separated quite easily, so clams that can be easily pulled open are dead.

Clams with cracked or broken shells die very quickly. Handle them gently to preserve shelf life.

Almost all commercial clam supplies come from the East Coast. Although there are many palatable species on the West Coast, especially in Washington and Alaska, the resources are either very small or are not heavily exploited. Alaskan resources overlap nursery areas for king crab, so clam fishing is prevented in order to protect the much more economically important crab stocks. Alaska decided in 1990 to encourage shellfish aquaculture; there is an impressive number of ventures being licenced and starting production.

Clams are subject to the controls of the National Shellfish Sanitation Program. All bags and containers must carry the license number of the shipper. If you open the containers, either to use the product or to repack, you must retain the tags for a minimum of 90 days (three months), in case there are problems with the clams later. If you repack clams into new containers, you must have a repacker's license from your state and put your own tags or labels, including your license number, on the containers you repack. The list of "State Officers Responsible for Issuing Interstate Shellfish Shipper Certificates" is contained in the monthly FDA publication, *Interstate Certified Shellfish Shippers List*. The information and the publication are available free from the FDA, HFF-344, 200 C Street, S.W., Washington, DC 20204. Most states require similar safeguards for shipments within the state, so you should check first, wherever you intend to ship. For more information on the NSSP, see the entry HANDLING SHELLFISH.

This section begins with a review of the terms used about clams, to make it easier to find particular information you want later.

Clam Glossary

ARK—Arks or arkshells comprise major groups of clam-like bivalves, but the term is not normally used in the United States.

BELLY CLAM—Softshell clams.

BUSHEL—Volume measure, equal to eight gallons. Live clams, in the shell, are normally sold by the bushel. Weight will vary considerably according to the type of clam, but a 60 pound bushel is about standard for hard-shell clams (quahogs) on the East Coast.

CHERRYSTONE—Quahogs counting 300 to 400 to a 60 pound bushel. Eaten raw on the half shell, though considered large for this purpose in New England. Also used for clams casino and similar cooked dishes. The distinction between cherrystones and topnecks as a size designation is rather confused.

CHOPPED CLAMS—Clam meat, ground up, for use in chowders, stuffed clams and other preparations. See below for more information.

CHOWDER CLAM—Large quahogs, up to 125 per 60 pound bushel. Used for baked stuffed clams. The meats are minced for chowders, clam cakes and similar preparations.

COCKLE—Small relatives of the clam used on the west coast for steaming. Cockles are an important European clam. There are vast supplies of blood cockles in Asia, though these may not legally enter the United States.

FRYERS—Softshell clam.

IPSWICH CLAMS—Softshell clams.

JUICE—Liquid byproduct from shucking and mincing clams. Used to add flavor to various clam dishes. Turns bad quickly unless pasteurized or preserved in some other way.

LITTLENECK—The smallest and most expensive grade of quahog, counting 450 to 600 per 60 pound bushel. Mainly used raw on the half shell, they are also sometimes steamed, although rather expensive for this purpose.

—West Coast hard shell clams, suitable for steaming but rather tough for half shell use.

MANTLE—The main meaty part of hard clams. Most of the rest is called the belly or stomach.

MINCED CLAMS—Same as chopped clams.

NECK—The siphons of softshell clams and geoducks. The necks look like miniature elephant trunks. Hard clams also have siphons, but they are much shorter and can be withdrawn fully into the shell. Softshell clams do not have room in the shell for the entire neck.

PUMPKIN—Very large quahogs, occasionally graded separately from regular chowder size.

SIPHON—See NECK, above.

SKIMMER—Surf clam.

STEAMER—In the northeast, usually softshell clams, which are known as steamers. Elsewhere, quahogs and manila clams are used for steaming and may be called steamers.

STRIPS—Product made by slicing the mantle of large clams. Usually served breaded, it may also be purchased unbreaded, in gallon containers, for breading by the end user.

STUFFED CLAMS—Stuffies. Real or imitation hard clam shell filled with clam stuffing and baked. Although these are frequently sold by the piece, they should be sold by the weight of filling, which can vary considerably and is not necessarily related to the size and weight of the shell.

TOPNECK—A fairly large grade of quahog, counting about 200 per 60 pound bushel. Many dealers do not bother to separate this grade from cherrystones and quahogs. They make a good uniform stuffed clam and are used in some markets for clams casino and similar recipes. Not all shippers grade topnecks separately but include them with cherrystones and chowder clams. Generally, you will be better off with a shipper that makes the distinction, since you want neither overlarge cherries nor small chowders. The shipper will be better off also, unless his practice has been to include all the topnecks with the cherrystones. Confusingly, some dealers grade topnecks smaller than cherrystones.

The most important clams used commercially in the United States are described individually in the following paragraphs.

SOFTSHELL CLAM: *Mya arenaria*. Also called Ipswich clam, belly clam, fryer or steamer.

These small, oval clams are found from North Carolina northwards into the Canadian arctic. There are small quantities, not usually

CLAMS

harvested, along the Pacific coast. It is also found, though rarely, in Europe. There are large resources, which are beginning to be more heavily exploited, in the Canadian Maritimes, but the greater part of U.S. domestic supply comes from Maryland and Maine. Maryland clams tend to be larger. Consumers and users vary in their preferences for large or small clams, liking what is usual in their locality. Canadian provinces set minimum sizes from 1.6 to 2.125 inches, while Maryland sets a minimum shell length of two inches (allowing five percent of the clams in a bushel container to be undersized).

Softshell clams are sold alive in bushels (normally packed in baskets to protect these delicate shellfish) or shucked in gallon containers. There is considerable labor cost in shucking and it is seldom if ever worth doing it yourself: the shucking plants in catching areas have expert labor capable of a very high output of clam meats.

Most of these clams are eaten either steamed, often as part of a New England clambake, or shucked, breaded and fried. People unaccustomed to them sometimes dislike the soft texture of the clam's belly, which is a major part of the animal. For this reason, clam strips (which are discussed below) have become popular throughout the United States as the generally acceptable form of fried clam. Softshell clams are generally too expensive to be used for chowders and other recipes requiring minced clams. They are not eaten raw, although you will sometimes find food writers referring to them when discussing raw clams.

Softshell clams are one of the most delicate of all molluscs to handle. The oval shells do not quite contain the entire animal so the clams dehydrate quickly, which kills them. It is vital to keep them in a moist atmosphere, using plenty of ice or wet seaweed. The shells are quite brittle, easily broken or chipped. Live clams in bushel baskets or bags must be treated very gently. Bangs, knocks and shaking can kill them. Although ice is essential for keeping clams cool and damp, never let them sit in ice: since the shells are not completely watertight, the clams are highly vulnerable to being killed by drip from fresh water ice. Dead clams must not be used.

Shucked clam meats are sold fresh in gallon cans. Clams have a strong odor, but it is easy to determine from the smell whether shucked clams are still in good condition. Most users prefer to buy their fryers already shucked and then bread the clams themselves. However, increasing quantities of fryers are sold already breaded and frozen, which is the most convenient way to handle them and avoids waste of a rather expensive product.

QUAHOG: *Mercenaria mercenaria*. Also called hard shell clam. This is the only hard shell clam sold fresh on the East Coast. Quahogs (there are various other spellings such as quahaug and quauhog) are an important seafood item, as they have been since native Americans used the shells as wampum (money). The southern quahog, *Mercenaria campechiensis*, is virtually identical for commercial purposes.

The definitions of the different sizes of quahog—littleneck, cherrystone, topneck, chowder and pumpkin—are shown in Table 1. These definitions are neither universal nor legally enforceable, but simply the best estimate, based on experience, of what you can expect to get. Different dealers have different grading ideas; many do not grade at all but simply ship the clams the way the fishermen supply them. Even the names are not applied universally. Cherrystone may be the smallest size in some parts of Connecticut. The topneck is a term used in New York and Philadelphia, but not much known elsewhere. Quahog is seldom used in New Jersey. Clearly, it is necessary to get to know your suppliers' definitions, or agree in advance what you expect to be shipped.

Note that there is a Pacific species known as a littleneck. Do not confuse this with the smallest size of quahog (see below). Some dealers may substitute small mahogany clams (see below) for littlenecks.

Note that the weight of a bushel of clams is not consistent and depends on the condition and size of the clams. A bushel bag containing 500 littlenecks might weigh as much as 60 pounds or as little as 45 pounds. It is preferable to use hinge width (the thickness of the clam across the two shells at the widest point) to define the sizes. Littlenecks will be over one inch, but not by much. Chowders will be over two inches. Quahogs will grow as large as six inches across the hinge. Some dealers now offer littlenecks as "count necks" and price them by the piece or by the bushel, containing a defined number of pieces. Since these small clams are generally sold and served by the half dozen or dozen, this seems a fair way to sell them.

Roughly speaking, quahog meats get tougher as the animals get larger. Small clams (littlenecks and cherrystones) are eaten on the half shell raw. Intermediate sizes may be baked or broiled. Large quahogs are minced and used for chowder, clam cakes and similar recipes.

Quahogs burrow into the mud bottoms of coastal bays and estuaries and are mostly captured by digging by hand with special long tools. Clam digging is a peculiarly hard and often dangerous way to make a

---

Table 1. A size grading for quahogs.

Littlenecks	450 to 600 per 60lb bushel
Cherrystones (or topnecks)	300 to 400 per 60lb bushel
Topnecks (or cherrystones)	about 200 per 60lb bushel
Chowders	fewer than 200 per 60lb bushel
Pumpkins or sharps	fewer than 80 per 60lb bushel

Chowders probably average about 125 in a 60lb container.

living, but it is part of the culture of coastal northeastern shore communities. Many fishermen would oppose newer or easier methods. Fishing methods are restricted by law to hand digging to conserve the resource and also to protect the sea bottom in which the clams live. Although powered dredges are readily available, they damage the habitat and would probably reduce future catches of clams. So far, only North Carolina permits mechanical dredging for quahogs.

Hogs are shipped live in bushel bags. The smaller the clam, the more costly the bushel. Large clams are shucked and sold as meats in gallon containers. Minced or chopped meats (the distinction is obscure but some suppliers insist there is a difference) are sold fresh in retail and institutional containers, though more often frozen. Clam juice is sold in plastic containers for prompt use and pasteurized in bottles and cans for long term storage.

Quahogs are found along the whole of the Atlantic coast from the Canadian Maritimes to the Gulf of Mexico, but they are not common further south, where the southern quahog becomes the more important species. This ranges from North Carolina to Texas. Most product comes from New York, Rhode Island, Massachusetts and Maine, but supplies from as far south as Florida are increasingly important. Production is related to the weather. Ice covering shallow bay areas prevents fishing; strong winds and high seas also keep the clammers ashore. Clams are therefore most abundantly supplied in the summer, when the weather is generally good.

Increasing numbers of local governments are planting small clams in protected areas to maintain and increase the supply. Clam culture technology is well developed, but commercially it is relatively undeveloped.

Opening clams requires a little practice and the proper tools. Clam knives are readily available as are inexpensive gadgets which hold the clam upright while forcing a small wedge between the shells. These are useful if you have a small volume of clams to open and no certainty of skilled labor. Opening large clams which will be used for cooking is best done by steaming them for a few minutes. This kills the clam, which causes the shells to open slightly, or at least makes the shells easily pried apart. If the steaming is done gently, no shells will be cracked or broken (which is important if you are making stuffies or other half shell items) and the meat will remain raw.

Although quahogs are hardy and will survive more poor handling than most other shellfish, they should still be treated gently. The shells can be cracked if bags are dropped or otherwise mishandled. Clams with cracked or broken shells should not be used. Bags should be stored in a cooler and kept damp. If displayed, they should be on trays or similar containers above the ice, not laid into ice.

Since clams are sold by the bushel, which is a volume measure and so has no consistent weight, you should occasionally check shipments against a correct volume measure. Many states have definitions or standard measures for oysters. The same measures will usually apply to clams. It is also important to check the accuracy and consistency of the size grading of quahogs smaller than chowders. This can be done by eye and hand. Count the number in a bushel, laying aside any that seem particularly large or particularly small. Then compare the weights of the largest and smallest. The largest should be no more than twice the size of the smallest. Ideally, the difference should be much less than this for littlenecks, where equal sizing is most important. Of course, if you have only 400 littlenecks in 60 pounds, you are getting cherrystones and should be charged accordingly, however consistent the grading.

Many popular clam products such as chopped clams and clam strips are sold frozen rather than fresh, since they deteriorate rapidly. Chopped clams and clam strips are more often made from surf clams or ocean quahogs than from the regular quahogs. These clams are dredged from deeper waters and are processed mechanically, giving a cheaper product that is suitable for further preparation.

SURF CLAMS: *Spisula solidissima*. Also called hen clam, sea clam, bar clam or skimmer. The American Fisheries Society, which is the primary authority on the names of marine creatures, calls this species the Atlantic surfclam (two words, not three).

This is a large hard shell clam dredged in much deeper water than quahogs. The mantle meat is used for clam strips and minced clams and is a vital foodservice product. Surf clams are found along the whole Atlantic coast. The clams are landed throughout the year, with peak production in spring and summer. They are subject to a Management Plan to control fishing and protect the resource.

Surf clams are only sold processed as strips, minced clams, or in the form of clam cakes, stuffed clams or other finished product. The meat is light colored and well flavored. Surf clams are an adequate alternative to quahogs for processed dishes.

OCEAN CLAM: *Arctica islandica*. Also called mahogany quahog, ocean quahog and black quahog. Mahogany clams look similar to quahogs, but have a brown or black periostracum (a hair-like covering) on the shell.

The meats tend to have an iodine taste and a dark brown color, which deterred the development of the resource. However, bleaching processes are used which improve flavor and color. Bellies are generally discarded, with only the mantle being processed, mainly for chopped clams and clam strips.

Although ocean hogs are tougher and darker than surf clams, they are an acceptable alternative at a lower price if properly processed. Dishonest dealers may occasionally substitute small mahogany clams for littleneck quahogs. This practice is not acceptable.

CLAMS

RAZOR CLAMS: *Siliqua patula.* The Pacific razor clam resembles an elongated softshell clam. They are found on exposed ocean beaches from the Aleutians to southern California, but the resources are generally small and fishing is restricted largely to recreational or subsistence harvesters. When available, meats are generally used in chowders or fritters. Live razor clams are very delicate and difficult to handle.

The Atlantic razor clam, *Ensis directus,* (the American Fisheries Society prefers it to be called the Atlantic jackknife clam) is also a very small resource, supporting insignificant commercial harvests.

MANILA CLAM: *Tapes philippinarum.* The American Fisheries Society prefers the name Japanese littleneck, but this would be commercially confusing, although the species is sometimes sold as a littleneck. Manila clam is a reasonable nomenclature which does not present any possibility of confusion with any other species. The scientific name has changed several times in recent years. You may find information on the species under *Venerupis philippinarum* and *Venerupis japonica.* These names are now considered incorrect.

Manila clams were accidentally introduced to the Pacific coast in the 1930s and now range from northern California to the northern edge of British Columbia. It is an important species in east Asia, where it is extensively cultured. It is being farmed now in Washington and British Columbia. Prospects for substantially increased supplies from aquaculture seem to be good.

As with the Pacific littleneck (see below), it is used for steaming. The species is marketed live and needs to be handled quickly. Note that it needs to be cooked for only about half as long as the littleneck, so the two species should not be mixed together in shipments. Unfortunately, they often are mixed. If you separate them, you will be able to supply your customers with much better product that will cook consistently.

PACIFIC LITTLENECK CLAM: *Protothaca staminea.* This is a small hard shell clam, similar to the manila clam, but with ribs radiating from the hinge. It is found on protected beaches from the Aleutians to Baja California. The species is usually sold with or alongside the manila clam. It should not be confused with the littleneck size grade of East Coast quahogs. It is generally considered too tough to eat raw.

BUTTER CLAM: *Saxidomus giganteus.* A large west coast hard clam found from the Aleutians to northern California. There are large resources, but because the species retains PSP toxins longer than most other shellfish, harvesting is restricted and landings are small. Butter clams from growing areas that are free of PSP are potentially an alternative to surf clam meat for chopped clams.

HORSE CLAM: *Tresus spp.* Also called gaper clam. These are very large clams found the length of the Pacific coast. They are known as gapers because the large siphon cannot be entirely enclosed by the shell. The shells are quite soft and brittle. They are mainly harvested incidentally by fishermen looking for geoducks. The resource is thought to be quite large.

GEODUCK: *Panopea abrupta.* Pronounced gooey-duck. This is the largest clam found in North America. The shell may grow to nine inches in length and the whole clam may weigh as much as nine pounds. Most commercially available geoducks are about three pounds. They are harvested mainly in Washington and British Columbia, although small numbers are found along the whole northern Pacific coast. Harvests are strictly limited by regulation. Some work is beginning to support possible farming of geoducks.

Geoducks do not live very long after they are removed from the beach, although they can be kept alive in holding tanks. The mantle may be used raw for sushi. The necks are skinned and made into steaks, or minced for fritters and chowder. Quality is related to color: the whiter the meat, the higher the price. Most of the limited catches are exported to Japan.

COCKLES: Cockles and arkshells are relatives of clams. The family includes the basket cockle, or Nuttall's cockle (*Clinocardium nuttallii*), which is thought to be a large, unexploited resource on the Pacific coast, especially in British Columbia. It can be used for steaming or other cooked clam recipes.

Huge quantities of small cockles are used in Europe and Asia. Blood cockles (which are arkshells) are so called because the meat turns reddish when cooked. These are important shellfish in southeast Asia. Although it is illegal to import clams from this region (see HANDLING), occasional quantities seem to be available in certain specialized markets. Because of sanitation problems, this practice is dangerous. Never buy any fresh or frozen clam product that does not have the proper tag, label and license number required by U.S. law.

RED CLAM: *Mactromeris polynyma.* Also known as Stimpson's surfclam, this is a small species from the Arctic. The resource extends into both Atlantic and Pacific seas. When cooked, the tongue or foot turns red, this characteristic makes it popular in Japan. Fisheries are developing in Canada, mainly targeting export markets, though small quantities are available in both the United States and Canada.

COBIA *Rachycentron canadum.*

Found along the Atlantic and in the Gulf, the cobia reaches 100 pounds, with many around 30 pounds. It is caught mainly by sport fishermen, but there is a ready commercial market when it is available. New regulations applying to federal waters limit even recreational fishermen to two fish over 33 inches daily.

Cobia is normally sold as skinless fillets (the skin is particularly tough), which can be sliced and used for frying, broiling or baking. Cobia is also sometimes smoked. It should not be called ling, which is a totally different fish. It should also not be called crabeater.

COD *Gadus morhua* is Atlantic cod or codfish. *Gadus macrocephalus* is Pacific cod or Alaska cod. The term true cod is not approved for Pacific cod (or for any other species, for that matter). There are several other cods, including tomcod and morid cod, none of which are significant commercially. Lingcod is not a cod (it is not a ling either). Black cod is an illegal name for sablefish.

Cod was historically the most important single commercial fish species in both the United States and Canada. It is still economically very important. It is a large, cold water fish, providing thick, white, meaty fillets which are low in fat and have a good shelf life. Atlantic cod was a mainstay of early North American settlements and still provides much of the income for communities in the northern Canadian Maritimes.

Atlantic cod ranges from North Carolina into the Arctic, on both sides of the Atlantic. The largest stocks in the western Atlantic are on the Georges Banks off Massachusetts and the Grand Banks off Newfoundland. Pacific cod comes from northern Pacific waters, along the Aleutian and Pribiloff Islands, as well as from Oregon, Washington, British Columbia, southeast Alaska and the Bering Sea.

Pacific cod is generally a smaller fish, caught further away from population centers than Atlantic cod. Consequently, it is most often used in frozen form. Although Pacific cod has a slightly higher percentage of water than Atlantic cod and is perhaps a little softer, the differences between Atlantic and Pacific cod for the end user are so slight as to be nonexistent. Pacific cod is now caught mainly by freezer trawlers that process the catch at sea, soon after it is netted. Fresh Pacific cod is mainly the product of local, inshore fisheries which are capable of offering excellent quality fish.

Atlantic cod of 100 to 200 pounds used to be caught. Fishing pressure on these older fish has had its effect and fish of 10 pounds are now regarded as quite large. Anything over 30 pounds is now unusual. Of domestic supplies, 85 percent is landed in Massachusetts. The Boston market grades dressed (head on) fish, as shown in Table 2. Most other places follow the same definition.

Fresh cod in the United States is invariably landed dressed with heads on. In Canada it is more often landed whole and round. Processing into fillets or steaks is done mostly at the port or by the distributor.

The usual product forms for fresh cod are as follows:

Table 2. Boston market size grades for cod.

small	under 1½ lbs
scrod	1½ to 2½ lbs
market	2½ to 10 lbs
large	10 to 25 lbs
jumbo, extra large or whales	25 lbs and over

DRESSED FISH, HEADS ON: This is how it comes from American fishing boats, in boxes of 120 pounds each. Because the weight of the fish together with the necessary ice makes an extremely cumbersome package to move around, this finds decreasing favor with distributors.

HEADLESS, DRESSED FISH: Rarely will dealers cut off the heads and ship fish in this form.

STEAKS: Slices cut vertically across the backbone. Fresh fish steaks are seldom graded, and in practice most are produced by the end user, who cuts dressed fish as he needs it. Shipping of steaks speeds deterioration of the fish because it exposes larger cut surfaces to air. It is preferable to ship dressed fish and cut the steaks at or close to the user's location.

FILLETS: These may be skin on or skinless. Skinless fillets are probably more popular. In New England the fish is skinned thinly to leave the layer of shiny fat flesh underneath, which is thought attractive for display of the fillets. An early sign of deterioration of the fillet is when the sheen fades and the fat appears gray rather than silver. Fillets are cut from graded whole fish and, if sold fresh, are usually not graded further by size.

BONELESS FILLETS: Generally V-cut (see FILLET), these may also be produced, but because the yield from the whole fish is reduced, tend to be expensive.

Fillets used to be packed in tins or tubs of 20 or 25 pounds. There are now many different types and sizes of packaging, competitively offering greater protection and shelf life and greater consumer appeal (when retail packs are involved). The trend to prepacks for retail use is growing and because it enables the port processor to control the quality throughout (and to take responsibility for his brand), it offers the consumer the best quality product and better assurance of consistency.

Fresh cod is imported in large quantities from Canada. Much of this moves through the Boston market structure and mixes with domestic fish, from which it is indistinguishable. Here again, Canadian packers are making efforts to prepack and brand their fish to give it better identity and reputation, with the intention of increasing their sales. Trade disputes between the United States and Canada have been caused by cod: the United States has claimed Canadian fishermen are subsidized, while Canadians have attempted to limit the export of whole fish, to keep processing work in Canada.

Fresh cod from Iceland and other European countries is sometimes imported by air. Whether the high quality of these products is worth the air freight is a question for the individual user to answer. Norwegian fish farmers are beginning to raise cod in sea pens, claiming a superlative quality product results. It is unclear whether many American consumers will be willing to pay more for top quality cod, whether farmed or wild.

Cod, like all fresh fish, should be kept well iced (see HANDLING). Quality of the whole or dressed fish is best indicated by the appearance of eyes and skin. Quality of fillets is indicated by the appearance and smell. For a full discussion of freshness, see QUALITY AND CONDITION OF FRESH FISH. Cod flesh consists of large flakes, joined together with connective tissue. If the flakes are beginning to separate, it is often a sign that the fish is aging. See GAPING for more details on this aspect. Most cod is caught in trawls, but line caught fish is thought to be better quality, as it may be alive when brought to the boat and can be gutted and bled at once. Line caught fish is available from small operators in Maine and Canada but most of it now sells along with the regular catch and is not distinguished. Pacific cod offered fresh in the northwest may be line caught (although some of the product displayed in retail stores is thawed, sea frozen fish).

Cod is used for smoking, salting and drying. Cod roes are a delicacy in many European countries, from Iceland to Greece. Cod cheeks and tongues are used in Canada and in parts of New England. There is a widely held belief that cod cheeks are used as a substitute for scallops. The cost of doing this would be excessive and any cod cheeks large enough are eaten by the fishermen, so the idea is basically impracticable.

CODEX ALIMENTARIUS

In English, this means "Rules for Foodstuffs." It is an internationally agreed set of standards, prepared under the auspices of two United Nations agencies, the Food and Agriculture Organization (FAO) and the World Health Organization (WHO). Many countries, including the United States, participate in the processes and use the Codex standards in their own regulations. There are numerous Standards and Codes of Practice for seafood products, ranging from canned Pacific salmon through frozen lobsters to fish sticks and portions. Codex standards aim at the production of clean, wholesome product. Applying them will mean your product is acceptable in a very large number of countries.

COOKING FISH

CONCH *Strombus gigas.* This species is also known as queen conch or pink conch. There are five other related species which are used interchangeably. Conch is a large, single shelled gastropod. The shells are decorative and prized as highly as the meat. The conch resources of southern Florida have been fished to exhaustion. Supplies come from the Bahamas and the Turks and Caicos Islands. The shellfish are caught by divers, but early efforts to farm the species are showing some encouraging results.

Raw, cleaned and skinned meats are offered fresh and frozen. They need to be pounded with a wooden mallet before use, to tenderize it. Make sure that the meat is fully cleaned. Avoid product with viscera remaining since this turns bad very quickly. Thin lipped conch, which is younger, is thought by some experts to be superior to thick lipped, or older, specimens. The younger shellfish are said to have whiter, more tender meat.

New England conch, frequently called scungili, is covered under the heading WHELK. The meat is rather different from that of conch. The two are alternatives in some uses. Squid can be used instead of conch in many recipes.

CONGER EEL *Conger oceanicus.* This is a large fish (up to 100 pounds) with firm, white, tasty flesh. It is not an eel, but the name is certainly not an advantage for marketing the fish in the United States. It is inexpensive and worth trying, if it should be offered.

COOKING FISH This is a brief note about basic rules for cooking seafoods. Too many people overcook fish and then complain about the result. The Canadian Cooking Theory, developed by the Department of Fisheries in Canada and extensively tested, works extremely well for all fish. (It does not apply to shellfish, but to all finfish.) It says that

> *Fish should be cooked for ten minutes per inch of thickness, measured at its thickest point.*

Baked fish should be cooked for 10 minutes per inch at 450°F. Broiling, pan frying, sautéing, poaching, steaming, even frying, all work the same way—ten minutes per inch. Rolled fillets are cooked according to the diameter (thickness) of the completed roll. When poaching or boiling, time the cooking from the moment the water returns to the boil after the fish is in.

Frozen fish can be cooked from the frozen state in exactly the same way, but requires double the cooking time, that is 20 minutes per inch. With thicker fish, over about two inches, it is preferable to allow the fish to thaw first and cook it as fresh. This Canadian Cooking Rule is very simple and very effective. It works for a thin flounder fillet as well as for a thick chunk of swordfish.

Microwave ovens work extremely well for fish, but the Canadian cooking rule does not apply. Retail packs increasingly include microwave instructions. The major requirement is to lay out the fish so that it is equally thick (or as near as possible); overlap pieces as required to achieve this. This helps to ensure even cooking.

Shellfish cookery requires even greater delicacy than fish cookery. Squid and shrimp can both be overcooked in three minutes. Oysters can be spoiled in two. Only whelk (scungili) benefits from lengthy cooking. Every other shellfish toughens more as it cooks longer. Microwaves can be used to open bivalve molluscs as well as to cook most shellfish.

CORAL
The roe of the lobster, while the eggs are still contained in the animal. When cooked, the coral turns red. It is well flavored and can be used to make sauces. It can, of course, be eaten with the rest of the meat of the lobster.

CORVINA
Cynoscion parvipinnis. A Pacific coast drum, prized as a sport fish but occasionally available through commercial channels. The fillets are very white and firm. It may sometimes be offered as seabass.

CRABS
The following crabs are important fresh in the United States and are discussed in order below: blue crab and soft shell crab; dungeness crab; rock crab; stone crab.

King crab and snow (tanner) crab are seldom available fresh. Almost all production is frozen. Both are discussed in detail in *The New Frozen Seafood Handbook,* available from Van Nostrand Reinhold/Osprey Books. Red crab, a deepwater East Coast species finding increasing acceptance, is also basically a frozen product.

BLUE CRAB: *Callinectes sapidus.* The newly moulted form is the soft shell crab. Sooks are mature females. Jimmies are males. Peelers are about to moult. Buckrams have already moulted and the shells are beginning to harden. There is an entire vocabulary just for the blue crab trade.

Blue crabs are caught from Cape Cod to Texas and are the most important crab (in volume terms) on the U.S. market. The Chesapeake Bay area is the largest source. North Carolina and Louisiana are also important producers. Gulf states have greatly increased their production in recent years. Crabs are also imported from Mexico. Blue crabs grow fast and are prolific, but their breeding grounds are in estuaries which are very vulnerable to pollution and to natural changes in the habitat. Consequently, the supply fluctuates widely from year to year.

The blue crab is a swimming crab, distinguished by the flattened, paddle shaped ends of its rear pair of legs, its blue tinted claws, which are approximately equal in size, and its oval upper shell, which extends at either end into long, sharp points. (Note that the blue crab, like all other crabs, turns red when cooked; the blue tint is only seen before the crab is cooked).

Blue crabs are available all year, but most landings are between June and October. Consumers tend to regard blue crabs as a seasonal item. Retailers sell them by the dozen. Soft shell crabs are available traditionally in June, July and August, but techniques have been developed which allow crabs to be held in tanks and persuaded to moult almost when required, so that the soft shell season has been greatly extended. The blue crab is most prized when it is in its soft shell phase, just after moulting (see below).

Live blue crabs are graded normally into three sizes, which are described differently by dealers: one person may use small, medium and large as descriptions. Another may call similar crabs medium, large and jumbo. Grading is done by eye and relates to the width of the shell across the top, from point to point.

The crabs are packed tightly into wooden bushel baskets, with the lid fastened on well. The idea is to prevent the crabs from moving around and fighting each other. If they are tightly packed, they cannot move. However, crabs at the bottom of the container may be crushed.

Live crabs should be kept cool, but not iced. A damp atmosphere helps to keep them alive much better than a dry one. Once a basket is opened, the crabs start to fight, which damages them. They also get turned over and do not survive long when they are upside down. Crabs shed legs and claws at will, so it is better to use tongs and pick them up by the body shell. Crabs picked up by a leg will often shed the leg. This is a defensive mechanism used in the wild to escape from predators, but consumers prefer to have crabs with all the legs and claws attached.

Crabs should be lively, but they will be less lively if they are cool. Warm but active crabs may live less time than cool and inactive ones. Experience with judging crabs is important. Although some wet seaweed in the top of the basket helps to keep the crabs damp, excessive packing material reduces the number of crabs received in the standard bushel basket.

Dead crabs should not be used, since the enzymes in the digestive system start to digest the crab when the animal dies and bacteria build up rapidly as the meat breaks down. Weak crabs that look as though they have little time left should be cooked immediately. Restaurants and retailers can both sell cooked crabs, which will keep for a couple of days if cooled promptly after cooking and kept well iced.

CRABS

Many consumers eat only the claws, discarding the body meat. Apart from being wasteful, this practice makes blue crabs expensive protein: if all the edible parts of the crab are eaten, blue crabs are quite inexpensive.

To cope with consumer resistance to preparing their own crabs, the industry has developed a substantial picking business, selling the cooked meat either fresh or pasteurized (and sometimes frozen, though freezing detracts from the flavor and texture of this product). Fresh crabmeat has a very short shelf life and, as with any cooked product, is highly susceptible to bacteria growth and contamination. Consequently, most fresh crabmeat is now shipped after being pasteurized, which gives a shelf life under refrigeration of at least six weeks. The meat is packed in 8, 12 and 16 ounce containers, usually graded into the following products:

Jumbo lump—The two large pieces of meat from the body. Body meat is white.
Backfin or lump—Should be the same as jumbo lump, but may contain smaller pieces.
Flake meat, special meat, regular meat or body meat—Smaller pieces of meat from the body.
Claw meat—Meat from the claws. This is brownish compared with the body meat.
Minced meat—Meat removed from the shell by a mechanical separator.
Cocktail claws—Claws with part of the shell left on, ready to eat. The remaining shell makes it easy for the diner to pick up the claw.

Crab meat should look and smell clean. A musty odor is a sign of decomposition. Bear in mind that parts of the crab meat vary in color from white to dark gray (in patches) and that yellowish marks may simply be traces of brown meat from the body of the crab. This brown meat is highly flavored but invariably discarded by American consumers, and so is not packed. However, small traces of it are unavoidable.

SOFT SHELL CRABS: These are blue crabs harvested just after they have shed one shell and before the new one hardens. (See MOULTING for discussion of this process.) Fishermen recognize when crabs are about to moult and also are able to keep crabs in cages in the sea, or in tanks on shore, until the right moment. The moulting process takes only a couple of hours and the crab must be removed from the water within one to two hours of shedding its shell, or the new shell will already be hardening and turning leathery—crabs in this condition are called paper shells or buckrams. A moulting crab will be eaten promptly by nearby hard crabs, so timing is important in all ways when producing soft shell crabs. The crabs are cleaned and then often frozen, since their shelf life is short. However, fresh soft shell crabs are generally available.

Although soft shell crabs appear to be very expensive compared with hard blue crabs, they contain many times as much edible material. They are popular with consumers and increasing in their appeal. Supply has been increased now that hard crabs can be kept in tanks and encouraged to shed their shells. Facilities to hold and handle the crabs are not very complex.

Soft shell crabs are graded according to the width of the top shell, as shown in Table 3.

Table 3. Size grades of soft shell crabs.

Whales, or slabs	over 5½ inches
Jumbos	5 to 5½ inches
Primes	4½ to 5 inches
Hotels	4 to 4½ inches
Mediums	3½ to 4 inches

These gradings are not universal. Different markets prefer different sizes. There is no quality implication whatsoever between the different sizes. State regulations governing minimum sizes vary. Maryland has a three inch minimum for soft shell crabs. Other states have larger size limits. Crabs missing legs and claws are considerably less valuable than complete crabs.

Clean soft shell crabs by removing the apron, which is the wide, triangular flap folded under the animal. Remove the gills, stomach and intestines from under the top shell and cut off the "face" behind the eyes. The rest of the crab, after washing and cooking, is edible.

Live soft shell crabs are very delicate and easily killed. Fresh cleaned crabs are susceptible to decomposition and should not be kept for more than a couple of days.

DUNGENESS CRAB: *Cancer magister.* This is one of the very best crabs to eat. Dungeness crabs are caught from California to Alaska. Oregon and Washington are generally the most important producing states. Commercial crabs generally are between one and a half and four lbs each. Seasons are specific, with most of the catch being taken in the late fall and early winter, but fresh crabs are generally available from some part of the crab's range during all months of the year. The heaviest landings are made when the season first opens. Catches vary greatly from year to year and from region to region.

Dungeness crabs are cooked as soon as they are caught and then distributed ready to use. Inshore crabs, especially from the San Francisco area, are known as bay crabs and are smaller than the ocean crabs. Fresh picked meat is also sold and is an excellent flavored product.

STONE CRAB: *Menippe mercenaria.* A small crab, mainly from the Gulf Coast of Florida. Only the claws are eaten. The rest of the crab is returned to the sea to regenerate claws for another fisherman—unless another marine predator eats it first. (Florida law prohibits the possession of whole stone crabs.) Claws are cooked and cracked at the knuckle and then shipped either fresh or frozen. Stone crab claws are an important and noted delicacy throughout the United States, although 80 percent of consumption is in Florida.

The name derives from the very hard shell, which requires hammering to crack it before the meat can be removed. The meat should come cleanly from the shell: if meat adheres strongly to the shell, the crab was not processed fast enough or it may have been frozen before being cooked. (Such claws are called stickers.) Cracks running around the shell indicate the claw may have been frozen. Stone crab claws have distinctive black patches on the ends of the claws and freckles on the shell. Lack of these identification features may indicate that some other crab— possibly an imported variety from South America—is being used.

Claws are graded as medium (under three ounces), large (three to five ounces) and jumbo (over five ounces). Some processors grade by counts: mediums six to eight per pound, large four to six per pound and jumbos one to three per pound. The minimum claw size allowed by law is two and three quarter inches in Florida and two and a half inches in Texas.

Fishing stone crabs is prohibited between May 15 and October 15 in Florida to protect the resource, which appears to be heavily utilized. Texas fishermen are permitted to catch stone crabs throughout the winter. The fishery in the western Gulf, which extends to Mexico, is comparatively new.

A Chilean crab, *Cancer edwardsii*, may occasionally be thawed and substituted. The shell is darker red and rougher, less shiny. Rock crab (see next paragraph) may also be substituted, but this is generally much smaller and has small ridges on the shell.

JONAH CRAB: *Cancer borealis*. This and the very similar rock crab are northeast species similar in shape to the Dungeness crab, but smaller. They are cheap because markets are undeveloped. Supplies are available mainly in spring and fall, between Massachusetts and Maryland. Many of the crabs are taken as incidental catches in lobster pots, and so are caught when most lobsters are being fished. The rock crab, *Cancer irroratus*, is very similar. It can be distinguished by the black tips on the claws and the scalloped upper shell.

Shoreside dealers can offer live crabs, kept in lobster tanks in the same way as lobsters are handled. Live crabs die faster than lobsters from the stresses of being transported, so inland sale of these crabs alive is not very feasible. Whole cooked crabs, properly packed in ice, survive for two or three days and are an excellent product. They are seldom over one and a half pounds.

Picked crab meat is a good, sweet product, though it tends to be in small pieces and shreds rather than lumps. It may be shipped fresh, frozen or pasteurized. It is, or should be, substantially cheaper than blue crab meat and can be used as an alternative ingredient. The crabs are picked by hand, which is expensive (it takes about a dozen crabs to yield one pound of meat). Development of picking machinery would certainly help to boost production, cut costs and enable more sales to be made.

Distributors could benefit from examining these crabs more closely, as they appear to offer high quality product rather inexpensively.

OTHER CRABS: Locally, many different crabs may be eaten, but the ones mentioned are the only species which are much used in fresh trade. As stated earlier, king, snow and red crab are invariably frozen items. Only very small amounts are sold fresh. As with all crabs, the flavor and texture of unfrozen product is definitely superior. Opportunities to use such product, provided it can be handled quickly enough, should not be overlooked.

The Pacific hair crab, *Erimacrus eisenbecki*, also known as the Korean crab (the FDA requires its meat to be called Korean variety crabmeat or Kegani crabmeat), which lives in the Bering Sea and grows to about two and one quarter pounds is thought to be another possible "new" crab resource. Hair crabs can be handled and shipped alive and this species sells for high prices in Japan. A similar species is found in Puget Sound. However, so far little is known about the abundance of these species.

CRAYFISH AND CRAWFISH

The name scientifically is crayfish, but, perhaps in deference to a Louisiana law calling these creatures crawfish, the American Fisheries Society has designated two important species as crawfish. These are the red swamp crawfish, *Procambarus clarkii*, which is now extensively farmed, and *Procambarus acutus*, the white river crawfish. In the United States, the other commercially available species is the signal crayfish, commonly called the Pacific crayfish, *Pacifastacus leniusculus*. Crawfish are clawed lobsters that live in fresh water. Mostly, they grow to only a few ounces, although there is a Tasmanian species that reaches eight pounds. North America has some 300 species of crayfish and Australia has another 100. Crawfish are highly regarded in northern Europe, where supplies of the native *Astacus astacus* has been hard hit by diseases. They are increasingly popular as they become more familiar and available in the United States.

Increasing quantities are being pond raised in Louisiana and adjacent states, sometimes as a joint crop with rice. Although these shellfish are found all over the country in lakes and streams, commercial production is limited mainly to the South and to Washington state and California. Improving technology and bigger acreages for farming crawfish should continue to ensure adequate supplies of this delicacy, which is successfully finding markets as production expands.

Farmed fish are available between about December and June, with peak supplies in April and May, when wild production is also available.

Crawfish are packed tightly and alive in onion bags for shipment. Bags are usually 30 to 50 pounds each. Crawfish will survive for three or four days in these bags if they are kept chilled with reasonable air circulation. Temperature should be close to 40°F. Some shippers offer purged fish which have been kept in tanks, without food, for several days so that body wastes are fully metabolized and excreted. This is a costly process, partly because crawfish are cannibals. The price reflects the additional cost and work.

Whole frozen crawfish are also available. Brine frozen product is preferable to blast frozen.

Cooked, ready to serve, whole crawfish can be purchased from some Louisiana processors. Shelf life is short, however, so the product must be shipped and used quickly. Peeled and deveined tail meat is also available fresh. This may be raw or cooked. Crawfish have quantities of fat in the head part, similar to the tomalley of a lobster. This is used as a basis for the rich sauces that are traditional with crawfish. Peeled crawfish can be bought with or without this additional item, although the inclusion of the fat shortens the shelf life of the product.

Peeled meats may seem expensive. Bear in mind that the yield from this animal is only in the region of 14 percent. Recipes requiring meats rather than the whole fish may be easier and cheaper to prepare if the processor has done the work of removing the shells.

Crawfish meat is soft and succulent, but has little real flavor. It carries well the rich sauces that are usually served with it.

The latest development in crawfish marketing is soft shell crawfish, which is finding ready acceptance in domestic and export markets. The stomachs and the glands in the head, called stones, are removed; the rest of the animal can be eaten, giving an excellent yield. The crawfish can be persuaded to moult readily in shedding tanks similar to those used with blue crabs to produce soft shell crabs. Most product is frozen, graded in counts of 21, 31 or 41 to the pound.

CREVALLE JACK
Caranx hippos. Also called crevalle, jack crevalle, toro, cavalla, horse crevally. Officially, it should be called either a jack or a crevalle. Although this game fish is found along the whole Atlantic coast, it is commercially used only in western Florida. Smaller dressed fish of one to three pounds are sold. Larger fish, ranging as big as 40 pounds, are caught, but the meat of the smaller ones is preferred. The fish is similar in shape to a pompano. The flesh is fairly dark and has reddish fat. Both color and flavor are better if the tail is cut off the

fish as soon as it is caught, so that it bleeds thoroughly. The larger fish are suitable for smoking and are sometimes filleted for retail sale.

The available resource is thought to be large. This is another inexpensive fish for which larger markets could perhaps be developed.

CROAKER *Micropogonias undulatus.* Also, regionally but not officially, called hardhead and grumbler. This is an often abundant, cheap, small panfish, normally about 12 to 24 ounces and landed in large quantities in the southeast year round, with peak supplies in spring. Landings (and availability of the fish in the sea) fluctuate enormously over lengthy cycles.

Chesapeake Bay production is year round. Atlantic Florida production is mainly in winter and spring; Gulf supplies are mainly summer and fall.

Basically a pan fish, tray packed in dressed form, croakers are also filleted or steaked. Markets extend from New York southward to the Gulf. The flesh is moist and well flavored.

CRUSTACEANS Shellfish with external skeletons and jointed legs. Although there are thousands of species, the commercially utilized crustaceans are very few—but very important, including shrimp, lobster, spiny lobster, crayfish and crabs. All of these decapod crustaceans (having ten legs). Nomenclature questions are best resolved by referring to the American Fisheries Society publication, *Common and Scientific Names of Aquatic Invertebrates from the United States and Canada: Decapod Crustaceans.*

CUSK *Brosme brosme.* A relative of the cod caught in small quantities in the colder parts of the North Atlantic. The flesh is white and firm and unfortunately not widely available. It is sometimes sold as ocean catfish, which has similarly good meat. The name tusk is sometimes incorrectly used.

CUTTLEFISH *Sepia spp.* Also called inkfish. Cepahalopod molluscs, similar to squid, often called sepia in the United States (sepia being the Italian name and also one of the scientific family names). Commercially offered cuttlefish is nearly always frozen and imported. It is used in many of the same ways as squid.

D

DEMERSAL Bottom feeding marine fish. The terms more often used in the United States are groundfish and bottomfish. Cod, flounder and similar fishes are all in this group.

DEPURATION The process of cleansing filter feeding molluscs to rid them of bacteria contamination. In the United States, it is not uncommon to relay these shellfish (primarily oysters, clams and mussels) in clean water so that they purge themselves naturally of bacteria and, possibly, other contaminants. Mechanical depuration under controlled conditions, in tanks on shore, is little used in the United States, although use is mandatory in many countries, including France, Spain and Australia.

In theory, depuration makes live shellfish safer to eat. The technique relies on the natural behavior of the animals: in the right conditions, they pump water through their digestive systems and extract their food from this water. Any bacteria in the animal is either digested or flushed out (over a period of a day or two) and, because the water used in a depuration plant is clean, no new bacteria can be absorbed by the shellfish. The water in a depuration plant is sanitized before the shellfish are placed in it. Sanitation techniques include:

- **Chlorine,** which kills the microorganisms in the water. The chlorine then must be removed from the water before the shellfish are added, or the shellfish will in turn be killed by the chlorine. This is a simple technology and well tried. It is used on a large scale in Spain, where all mussels, clams and oysters must pass through depuration facilities before they can be marketed.
- **Ozone,** which similarly kills microorganisms, but dissipates on its own very quickly from the water. Ozone depuration plants are used in

France, which also requires depuration of the large quantities of all live filter feeders consumed.
- **Ultraviolet** light, which again kills microorganisms. In UV systems the water is pumped continuously past a bank of lights, which kill bacteria in the water as they are pumped out of the shellfish. This technology is the one most likely to be used in the United States, although in practice it requires careful management of many factors, including the water flow rate, the turbidity and temperature of the water and the condition of the ultraviolet light sources.

Depuration can be used to add an extra margin of safety to shellfish harvested from approved waters. In certain circumstances, the techniques can be used to make it possible to harvest shellfish from slightly less clean areas. Although relaying is usually used in this circumstance, mechanical depuration can also be used. Depuration techniques cannot be used to purify shellfish that are heavily contaminated, because the time taken to cleanse viruses from the animals can be so long that they would die of starvation before they were sanitized. (Adding algal feed to depuration units is far from being economically feasible.)

Any use of depuration in the United States is subject to the controls of the NATIONAL SHELLFISH SANITATION PROGRAM (NSSP). See the entry under that heading for more information. The FDA, which administers the program federally, has not encouraged mechanical depuration.

The reasons for caution about depuration include the following:

- Supervision has proven difficult. New Jersey closed its last commercial depuration plant partly because authorities found it too difficult and expensive to monitor it for proper procedures.
- Operating a depuration plant requires considerable management effort to ensure that all of the factors involved in successfully cleansing shellfish are constantly monitored and controlled.
- An incorrectly operated plant can actually add to the bacterial load in a batch of shellfish.
- The technique is effective in removing certain bacteria. Other bacteria and viruses are not reduced sufficiently, partly because viruses may migrate from the gut into the flesh of the shellfish, where depuration cannot reach them. Consequently, the fact that shellfish have been cleansed in a depuration plant may give consumers a false sense of security.
- Tests to determine whether viruses are present in shellfish are lengthy and costly. They cannot usually be performed on shellfish that are intended to be marketed alive. The NSSP monitors growing

waters for viruses; shellfish harvested from approved waters should be free of them.
- If depuration is widely seen as a way to permit the marketing of shellfish from contaminated waters, efforts to clean up shellfish growing areas may be reduced. These efforts are already inadequate, with many states reporting reduced areas of approved waters over the last few years, despite a universal intention to clean up the marine environment.

Of course, a great deal of the risk from eating raw shellfish could be reduced if the animals were frozen. Most bacteria are killed if held below 10°F for several days. Oysters and clams freeze very well. Few people can tell whether a raw oyster or clam was previously frozen. However, there is enormous prejudice against freezing these products.

For more information on this topic, see this author's *Shellfish—a Guide to Oysters, Mussels, Scallops, Clams and Similar Products for the Commercial User*, published by Van Nostrand Reinhold/Osprey Books.

DIPS A number of phosphate chemicals are frequently used to limit drip loss and whiten fish and shellfish. Sodium tripolyphosphate, potassium pyrophosphate and sodium pyrophosphate are some of the chemicals used. Others employed include potassium tripolyphosphate and sodium hexametaphosphate. Each has certain advantages and certain disadvantages. Sodium tripolyphosphate, for example, reacts with fats in the fish and produces soap. It is also difficult to dissolve. Many commercially available dips contain mixtures designed to reduce problems and make the product easy and effective to use.

Generally, the product is dipped for a short time in a solution of the chemical mixed with water. The effect is to reduce the natural loss of water from the cells of the flesh. This has obvious advantages throughout the distribution chain, since the original weights shipped from the processing plant are more or less maintained in the package.

It is particularly important with scallops to prevent drip loss, since this shellfish will continue to drip quite rapidly and can lose a large percentage of its original weight. Scallops tend to darken as they age and the dips also help to prevent this from happening. Dips whiten the flesh of fillets as well as of scallops, which is important with gray colored fillets such as Atlantic pollock. Dips help to reduce gaping of the muscle.

Properly used, dips make an important contribution to maintaining original weights and improving the appearance of seafoods. So far as is known, they are harmless. However, some dips (not all) add significant amounts of sodium to the food, which could detract from the spreading uses of fish in popular low sodium diets.

Phosphates are easy to use. The correct concentrations are clearly marked on the labels. If too much is used, it can make product tough when cooked, sometimes so tough that it appears never to cook at all. This problem is not encountered often, and relates mostly to imported frozen product, especially peeled shrimp and scallops. It can, of course, happen by accident. It is seldom possible to spot excessive phosphates until the product has been cooked, when inevitably it is a difficult problem to resolve.

Freshwater fish may suffer from overdoses of phosphates if the processor is inexperienced. Some fresh water species absorb enormous amounts of the chemicals, apparently because of the nature of the flesh.

Some dipped product foams when washed. This may be due to protein from the flesh, or to the reaction of the fat in the flesh with the alkaline dip, which actually makes soap. Usually, substantial foaming is a sign that far too much dip has been used.

The major problem of misusing dips on fresh fillets is the application of heavy concentrations to disguise product that is going bad. It is possible to dip fillets so that off odors are concealed until the product is cooked, when the smell becomes all too apparent. By this time, of course, the fillet is in someone's kitchen and the perpetrator may often avoid retribution, at the cost of damaging the reputation and market of the industry as a whole and of the local supplier in particular.

The other problem with dips is that they can sometimes be misused to increase the weight of product, rather than to prevent the weight declining. Scallops, which can absorb water, if soaked rather than dipped in phosphates, will actually increase in weight and will retain some of that weight during shipment.

The use of dips has to be noted on labels as an ingredient. This rule is widely ignored by the seafood business, as it is by other segments of the food industry where dips are also enthusiastically used and abused.

Some dips are available with additional ingredients, such as citric acid and sorbic acid. Sorbic acid reduces the surface bacteria which cause spoilage. Citric acid slows the growth of the remaining bacteria by reducing the pH of the surface of the flesh, thus making it a less hospitable environment for bacteria. Citric acid also helps to whiten the flesh. These ingredients are classified by the FDA as Generally Recognized as Safe. See also ADDITIVES.

DOLPHIN The dolphin fish is described in the item MAHIMAHI. The mammal dolphin or porpoise is a protected animal and not used. American tuna canners go to considerable lengths to ensure that dolphins are not accidentally caught with their tuna. The United States embargoes tuna from countries that do not make similar efforts. There is confusion when mahimahi is described as dolphin or dolphin-fish. There is absolutely no relationship or connection or similarity between them. If you offer mahi as dolphin, there is always someone who thinks you are frying a protected animal. It is much better to stick to the Hawaiian name mahimahi.

DORIES The American John Dory is *Zenopsis ocellata*. This is very similar to the European John Dory, *Zeus faber*. Occasionally offered fresh, these are small fish rarely exceeding three pounds. The meat is white, firm and sweet. This is the fish called St. Pierre (St. Peter's fish) in French.

Smooth oreo dory, *Pseudocyttus maculatus*, is available frozen from the southern hemisphere. New Zealand pioneered the fishery for this species, which provides firm, white fillets. The fish grows to about two pounds, giving fillets about the same size and general appearance as those from orange roughy. There have been incidents of substitution of the cheaper dory for orange roughy. Because of the increasing use of air freight for fresh seafoods, it is possible that oreo dory may be offered fresh.

Black oreo dory, *Allocyttus spp.*, is a smaller southern oceans fish, producing smaller, thinner fillets of under four ounces. This is less expensive than the smooth oreo and less likely to be offered in chilled state.

DRAWN Eviscerated, that is, the guts have been removed.

DRESSED Dressed fish is gutted and usually scaled. Dressed fish may also be split, ready for salting. Dressed crab is cooked and cleaned. The meat is removed, seasoned or mixed with mayonnaise and sometimes other ingredients. Then it is placed back inside the shell of the crab and carefully displayed. Dressed crab should be completely ready to eat, needing only a diner with a fork. It does not have much shelf life, so it should be prepared, if possible, immediately before sale.

DRUM, BLACK *Pogonias cromis*. A Gulf and Atlantic species which currently comes mainly from Texas and Louisiana. The species is largely gone from the middle Atlantic region, apart from small landings in spring and early summer.

Fish are sold dressed and filleted, sometimes frozen. Usual commercial size is about four to five pounds, though much larger fish are caught by sport fishermen. The flesh is rather coarse, especially from larger fish. These are mainly used for chowders. Smaller fish are popular in

Chinese cooking and also may be used as an alternative to redfish, although it is not such a good fish. Like the redfish, black drum may contain trematode parasites, so it should be cooked thoroughly. This fish must be bled immediately on capture to lighten the flesh and reduce parasites.

Drums include a large group of quite different fish, some of which are used in the seafood trade. See BASS, CORVINA, CROAKER, REDFISH, SHEEPSHEAD and WEAKFISH (sea trout) for other drums.

E

EEL *Anguilla rostrata.* The FDA prefers this species to be called freshwater eel or American eel. The American eel is very similar to the European eel (*Anguilla anguilla*). The major difference is that the American eel has seven or eight fewer vertebrae. The two species both spawn in the Sargasso Sea area of the Atlantic. The Japanese eel, *Anguilla japonicus,* is also similar.

Eels are snake like in shape. This characteristic has prevented many people from sampling them. Like squid, this item may need a different name if it is to secure consumer acceptance.

Eels are catadromous, meaning they spawn at sea and return to fresh water to grow, the opposite of salmon, river herring and most other fish which spend part of their lives in fresh water and part in the sea. U.S. production is small and much of it is exported since domestic markets are restricted to a few of the largest cities and to a number of shoreline areas where people appreciate the fine eating qualities. Italian communities in particular cook and eat eels at Christmas and this constitutes the major part of the domestic market for unprocessed eels. It is believed that eel resources could support enormous expansion of fishing.

Most of the small U.S. production is from the middle Atlantic states, especially Virginia. Brown (or yellow) eels are the main catch: this is the juvenile stage when the fish is migrating to fresh water to grow. They are usually between one and three pounds. Silver eels, which are the same species returning to the sea to spawn, may be as large as 20 pounds.

Very small eels, up to three inches long, are called glass eels or elvers and are a delicacy in some Asian countries. Elvers are also exported to Europe and to parts of Asia to be grown and fattened in aquaculture ponds for later marketing. Aquaculture now produces most of the world's eel supply. Japan, Taiwan and China are all major producers, selling mostly to the Japanese market, where eels are an important food.

Fresh eels are usually shipped alive in tanks specially equipped with air pumps to keep the water full of oxygen during transit. They are usually purged, by being held in tanks without food for 24 hours.

Eel flesh is extremely palatable. The meat has fairly high fat content, but is white and firm. Eels may be stewed, boiled, broiled, fried—almost any method of preparation works well. Eels may be filleted, but if they are cooked with the bone the flesh is easy to remove and there are no small bones to complicate the operation. Eels must be skinned before use. Smoked eels are a considerable delicacy.

It is unfortunate that prejudices prevent people from trying eels, which are an excellent seafood and sadly neglected.

LAMPREY: *Petromyzon marinus*. Lampreys are not eels and are quite different from eels. They have cartilage, not bones. In past centuries they were thought to be eels and were also considered great delicacies. Sea lampreys have infested the Great Lakes, where they kill other fish. They are also found in the Atlantic and are a potential resource in the Columbia River. No one in the United States eats them any more and most people are surprised to learn that they are edible. They can be used in many eel recipes.

F

FAT FISH Those with the fat distributed through the muscle, such as herring, salmon, mackerel and tuna. Other fish such as cod, haddock and flounder have most of the fat in the liver and so have less oil in the meat. In general, but not always, fat fish have darker flesh. In almost all fish, the fattest parts are the belly flaps and along the lateral line, where there may be a strip of darker meat.

FILLET U.S. Grade Standards define fillets as follows:

> Slices of practically boneless fish flesh of irregular size and shape, which are removed from the carcass by cuts made parallel to the backbone, and sections of such fillets cut so as to facilitate packing.

Many fillets therefore have a few bones. Boneless fillets are cut to remove the remainder of the bones. See BONELESS FILLETS for the definition.

Fillets should be trimmed so that they are free of napes, belly bones, membranes and fins. J-cut and V-cut fillets are cut in the described shapes (at the nape end) in order to remove the pinbones and produce boneless fillets. Flatfish fillets should be thoroughly trimmed of the frill, which is the bony part of the fin surrounding most of the fish.

FILTER FEEDERS Certain molluscs, such as oysters, scallops, clams and mussels, pump seawater through their systems and filter from the water the nutrients they need. See DEPURATION and SHELLFISH SANITATION.

FINNAN Finnan haddock, finnan haddie. Originally, finnans were haddock split with the backbone left attached to one side and then hot smoked. The name is now frequently applied to any similar fish which has been smoked, with or without the bones. American consumers prefer boneless fish. Cod, pollock and whiting are used at least as much as haddock. See SMOKED FISH.

FISHERIES MANAGEMENT
National objectives to control the catch of a marine species in order to maximize the resource are administered by eight regional Fishery Management Councils authorized under the Magnuson Act. The councils devise management plans covering federally supervised fishing areas (in general from three to 200 miles from the coast). States participate in the councils and generally, but not always, administer their own territorial seas to conform with federal plans.

Because the managers seldom have sufficient data to know precisely what is happening to a fish resource, the management plans tend to be biased in favor of restricting fishing. Further, recreational fishing interests have a great deal of political influence and have successfully argued for restricting certain commercial fisheries. Consequently, some of the increases in commercial supplies promised when the system was set up have not materialized.

It is important to monitor existing and proposed plans. They can have very substantial effects on both short term and long term supplies of certain seafoods.

FISHING METHODS
The way a fish (or shellfish) is harvested may have a bearing on product quality. There are literally thousands of ways of catching fish (see a very large FAO book called *Fishing Gear of the World*), but commercial supplies are mainly produced by some form of trawling, seining or gillnetting. All three of these industrial techniques use nets. Some commercial species are caught with hooks and lines: salmon, swordfish and shark frequently, sablefish and even cod often enough to be notable. Other fish are harpooned (swordfish, again) and still others are trapped (squid and scup in Narragansett Bay, for example).

The catching technique used may affect the quality of the fish when used. This is not a simple matter and there are numerous considerations involved such as the species caught, the local air and water temperatures when it was caught and the physical condition of the fish. In general, catching technique has less of an effect on quality than does the way the fish is handled immediately after capture. But fishing technique does affect that in some ways, too.

Butterfish, scup and squid, for example, are worth more if they are caught in traps rather than nets. Trapped product can be handled gently. There is less scale loss or skin abrasion, less bruising and generally less damage likely. Nets, especially large ones, cause damage to fish that comes out of the water at the bottom of the net, partly because of the heavy weight of the other fish, partly because of the roughness of the net itself. Herring taken in large mid-water trawls are more likely to be damaged than herring taken in small gillnets.

Netted sharks may suffocate because they have to keep moving to force water through their gills to supply oxygen. Sharks taken on hooks are more likely to keep moving and come out of the water alive. Since sharks must be bled as soon as possible to reduce the risk of ammonia building up in the flesh, this can be important. With other species, though, gillnets may produce a better product than lines. When fish fight on a hook, they use up stored glycogen. The result is a tougher, less sweet meat. the same thing happens if fish are left to flap around on deck; most species should be killed (and bled) as soon as they are landed.

Excellent mackerel can be landed with hooks, but purse seining is far more efficient and the market differential seldom justifies the additional cost. On the other hand, giant bluefin would probably be unsalable unless hooked, because of the special quality requirements of the Japanese market, which buys most of the world catch.

Troll caught salmon, hooked and handled individually, is generally superior to netted salmon. Note that this is only partly because the hooked fish is handled individually and not bruised in the net. It is also because the slower catch rate means that the fishermen gut the trolled fish. Netted salmon is nearly always taken to shore whole and then dressed.

In other words, the way the fish is handled on the boat can be at least as important as the type of gear used to land it. Bleeding, careful handling and, above all, rapid and adequate icing will all contribute greatly to high quality product. Fish that sits in the sun on deck for an hour, however carefully caught, will lose enough quality to ensure that it will never be a first class product. Overall, it is better to pay attention to the quality of the product rather than making selections based on fishing methods.

FLAKE Many fillets are composed of flakes of flesh, joined together with thin connective threads or tissues which run through each flake and join the shiny membrane that separates each of the flakes. A cod fillet from the full length of the fish has about 50 flakes. Fish with large flakes, such as cod and haddock, are often preferred over fish with small flakes. See GAPING for more information.

FLAKE MEAT: Also called special meat and body meat. Meat from the body of the blue crab. A less expensive grade than lump meat, it consists of smaller pieces from the body.

FLETCH A boneless fillet from halibut. The term is sometimes applied to large pieces from swordfish, shark and other large fish.

FLOUNDER AND SOLE This heading deals with sole and flounder together. They are both covered under flounder not because this is considered a preferable name, but simply because it comes first in alphabetical order. Flatfish consist of numerous species that live on the sea floor and have both eyes on the same side of the head. Hold a flatfish with its belly towards you and the eyes on top. Some species have the head and eyes at the left and are known as left-eyed flatfish. Others have the head to your right and are called right-eyed flatfish. If you have both a bottom and a top fillet you can reconstruct the fish sufficiently to be able to work out which type it is. The distinction is sometimes important for identification.

Flatfish include flounders, soles, turbots and halibuts. Most flounders and soles are quite small, giving fillets between about three and 16 ounces. Flatfish fillets are invariably boneless, because the fish do not have the equivalent of pinbones and the frill of small bones around the edge of the fillet, which is the fin, is easily removed when the fillet is trimmed. Bonelessness and bland taste together account for much of the popularity of these fish.

Fresh flounder fillets are usually packed in tubs, tins or boxes, or they may be packed in the newer vacuum sealed or plastic containers. Fillets should be fairly uniform in size, white and almost odorless. Skins are normally removed, although a few processors may leave white skins on the fillets. (Many species of flounder and sole have white skin on the underside and dark or patterned skin on the upper side, which camouflages the fish while it is alive.) The subcutaneous fat, which is silvery, is normally left on the fish as it provides an attractive appearance for display. If this small patch of fat is dull or gray, the fillets are beginning to age. Old fillets may start to gape, but since the flake is small, they appear to be falling apart rather than gaping. Fillets should be cleanly cut—they should not have ragged edges and there should be no traces of viscera on the flesh. Flounder and sole fillets are delicate, partly because they are thin. They should be handled with care.

Flounder and sole may also be shipped simply dressed. The small gut cavity should be clean. Heads may also sometimes be cut off. In this form, the smaller fish are used for grilling, baking and even stuffing. Small fish are often shipped without being filleted because the yield of flesh from a small flatfish is too small to be economic. If you are buying fish for filleting yourself, bear this in mind.

Also bear in mind that fish that has just spawned or that is about to spawn may be in very poor condition. Fish about to spawn may be shipped with the roe left in the gut cavity. Since few people will eat this rather oily tasting roe, it is additional waste for the user. Fish near or just after spawning time is usually watery as much of the fish's metabolism has been devoted to producing roe or milt rather than maintaining muscle quality.

FLOUNDER AND SOLE

Like other fish, flounder and sole should be kept well iced, with the ice changed frequently enough to avoid cavitation. Very low in fat, the flesh keeps well if properly handled.

Flounder and sole nomenclature is simpler than it used to be, since the FDA published its *Fish List* allowing many species to be called by either name. It is still complex enough. The basic distinctions, enshrined in the U.S. Commerce Department section of the Federal Code, are shown in Table 4. Note that even here, *Pseudopleuronectes americanus* changes from a blackback flounder to a lemon sole when it grows beyond 3 pounds. The FDA rejects the term blackback altogether, preferring to call the species either winter flounder or lemon sole.

The following comments on some of the more important species include the FDA's preferred market and common names. Note that the market names include 40 flounder and 25 sole. There are a further three plaice, one sanddab, one tonguesole and seven different turbots. Bear in mind that world wide there are very many more species, some of which might well be offered for sale in the United States. The *Fish List* has been edited and reproduced in the Appendix for easy reference.

ARROWTOOTH FLOUNDER: *Atheresthes stomias*. No other names are permitted by the FDA; arrowtooth sole is illegal nomenclature. A large flounder found from Oregon northwards through Alaska and around to Japan. The flesh is extremely soft and falls apart unless the fish is treated and cooked with great delicacy. The problem is a myxosporidian

Table 4. NMFS definitions of flounder and sole names in the United States.

Common name	Scientific name	Coast
	SOLE	
Dover sole	*Microstomus pacificus*	P
English sole	*Parophrys vetulus*	P
Gray sole	*Glyptocephalus cynoglossus*	A
Petrale sole	*Eopsetta jordani*	P
Lemon sole	*Pseudopleuronectes americanus*, over 3½ lb	A
Rock sole	*Lepidopsetta bilineata*	P
Sand sole	*Psettichthys melanosticus*	A
	FLOUNDER	
Blackback	*Pseudopleuronectes americanus*, under 3½ lb	A
Yellowtail	*Limanda ferruginea*	A
Dab, plaice	*Hippoglossoides platessoides*	A
Fluke	*Paralichthys dentatus*	A
Starry flounder	*Platichthys stellatus*	P

Source: Federal Code, 50 CFR 263.20

parasite which produces an enzyme which breaks down the flesh. Technologists are working on ways to process and package the fish to minimize this problem.

Canadian fishermen call this fish turbot. Fillets between four ounces and two pounds are available. Although most arrowtooth is imported from Japan and Korea frozen, some fresh fish is available and generally should be avoided.

DAB: *Hippoglossoides platessoides.* This species is also called plaice and American plaice (or, in Canada, Canadian plaice). In international statistics it appears as long rough dab. This is a right-eyed North Atlantic species generally around one to three pounds, although it grows as large as 12 pounds. There is a similar Pacific species called Bering Sea flounder, *Hippoglossoides robustus.* Maine and Massachusetts produce nearly all the domestic dab supplies.

The flesh is quite firm and has good flavor. Dab is a good quality flatfish with a wide market.

DOVER SOLE, GENUINE: *Solea solea* or *Solea vulgaris.* The original "genuine sole" of Europe. This is a very expensive flatfish with firm, sweet flesh which is very white when cooked. Very small quantities are flown in from France and Holland from time to time, but most U.S. supplies are frozen. Dover sole should be cooked on the bone for best flavor and yield, so fish should be purchased dressed. Dover sole is the basis and supreme raw material of the best French fish cookery. However, it is so expensive that few people will be willing to pay for the genuine article.

DOVER SOLE: *Microstomus pacificus.* The FDA mandates the name dover sole, in this case not shortened simply to sole. A Pacific right-eyed species, dover sole is variously described as very poor, or good eating. Possibly the distinction is between handling and actual taste. The fish has a great deal of natural skin slime and is still sometimes handled unwillingly because of this. Fishermen sometimes call it slime sole. Once skinned and filleted, however, it is acceptable but inferior to petrale and English sole among the Pacific species.

Dover sole is caught from northern Alaska to southern California. It is a fairly deep water fish and grows to around 10 pounds. It is now one of the more important west coast flatfish. Alaskan fish is often better than Californian dover sole, which has a tendency to jelly. The species generally does not keep well, so it is better distributed in frozen form.

ENGLISH SOLE: *Parophrys vetulus.* The FDA permits this species to be called sole or English sole. In California it may be called lemon sole, which is also the preferred name in Canada. It is quite unlike any sole likely to be found in England. It is also unlike the East Coast lemon sole or the European lemon sole, which are two different flatfish. A small right-eyed flatfish from the full length of the West Coast, it generally reaches about three pounds. The flesh is delicate and well regarded.

FLUKE: *Paralichthys dentatus*. Summer flounder is the alternative name for this large North Atlantic left-eyed flatfish with excellent, firm, white flesh. Fish up to 10 pounds are common, and much larger flukes are taken occasionally. The market for fluke is particularly strong in the Middle Atlantic states (where the species also supports an active sport fishery).

Most fluke comes from the middle Atlantic states and Rhode Island. Regulators believe that the species is being exploited as fully as possible, which indicates supplies are unlikely to increase. Management plans and state regulations are in place to restrict production. July to September sees the greatest landings in the northern and central parts of the fish's range; November and December are peak months further south.

The flesh is firm and white and widely acceptable to consumers. Some fish produce fillets that are large enough to be cut in half along the lateral line of the fish and sold in this form. Fillets are sold both with and without the skin. The species is highly regarded in Japan, where it is eaten raw. Fish for this purpose may be shipped from the United States by air.

GRAY SOLE: *Glyptocephalus cynoglossus*. According to the FDA, this species may also be called witch, witch flounder, flounder or witch sole. It is a right-eyed flatfish, normally the most expensive of the Atlantic soles. The gray sole fillet is rather long and thin, and has a particularly good flavor, which presumably accounts for the substantial price premium paid. It is found on both sides of the North Atlantic and frozen fillets are imported. Maine and Massachusetts supply most of the domestic production; Canada supplies a good part of U.S. supplies of this species. Gray sole grows to about four pounds, giving fillets of around eight ounces.

PETRALE SOLE: *Eopsetta jordani*. This is a right-eyed flatfish that may be called sole or flounder. Commercially the most important west coast sole, petrale is found from southern California to northern Alaska. It grows to about seven pounds and has good, quite firm flesh. It is usually regarded as the best eating of all the West Coast soles.

REX SOLE: *Glyptocephalus zachirus*. This right-eyed species may be called sole or flounder, but should not be called long finned sole. A small west coast flatfish, reaching about two pounds and found the entire length of the North American coast. It is closely related to the gray sole, and is similarly elongated in shape. Also like the gray sole, the fillet is thin and the flesh very delicate. Because it is small, rex sole is difficult to fillet and is frequently sold whole, though Alaskan fish tend to be large enough to fillet.

ROCK SOLE: *Lepidopsetta bilineata*. This species may be called sole or flounder, but should not be called roughback. Found from the Bering Sea to California, it is a right-eyed flatfish caught in quantity in Japan. Rock sole is fairly small, seldom reaching five pounds. The main resource for U.S. fishermen is off California. It is also the most important Canadian west coast flounder/sole. The creamy flesh is firm. Resources in Alaska are being developed.

SAND SOLE: *Psettichthys melanosticus*. This is a right-eyed flatfish similar to the petrale, found the full length of the West Coast. It may be called sole or flounder.

In Europe, there is a different fish called sand sole which is similar to the Dover sole or true sole.

SOUTHERN FLOUNDER: *Paralichthys lethostigma*. The FDA permits this species to be called fluke. There is a very similar Gulf flounder, *P. albigutta*, which is rarely distinguished commercially. Left-eyed flatfish similar to the fluke, the southern and Gulf flounders rarely exceed three pounds and are usually around two pounds. They are caught throughout the Gulf coast and along the Atlantic coast from North Carolina southward. Florida and Alabama produce over 80 percent of the landings. Availability is year-round, with highest landings in November and December in the Gulf and in January from South Carolina. Fresh supplies are therefore most readily available at the season when the North Atlantic flounders are least abundant. The fillets are good quality, similar to the fluke. Most supplies are sold as skinless fillets, fresh.

STARRY FLOUNDER: *Platichthys stellatus*. An important west coast flounder, found throughout the North Pacific, from California to Japan. The flesh is firm and well flavored. The species is economically important. Fish grow to over ten pounds, though most sold commercially are under three pounds.

WINTER FLOUNDER/LEMON SOLE: *Pseudopleuronectes americanus*. The FDA permits winter flounder but not the commonly used term blackback flounder. This is another right-eyed flatfish and is the second most important commercial flounder species of the Atlantic coast. Over 90 percent of domestic supplies come from New York, Rhode Island and Massachusetts. The fish is found inshore during the winter (hence the name winter flounder) and migrate to the offshore banks for

FLOUNDER AND SOLE

the rest of the year. Most landings are in April, May, October, November and December.

The species is popular in the midwest and the northeast. Small quantities are frozen, and more is imported frozen from Canada, but the bulk of supplies are distributed fresh. The fillets are delicate but the flavor is good. Spawning fish, which come inshore in the winter, have very soft flesh and should be avoided. The fillets from the side with dark skin tend to be a little gray, but turn white when cooked.

Frozen lemon sole from the Far East is sometimes available. These fillets may be very thin and it is not the same fish as the Atlantic lemon sole.

YELLOWFIN SOLE: *Limanda aspera.* Another fish that may be called either flounder or sole, this is generally a small fish, between four and 24 ounces, found in the Bering Sea and throughout the North Pacific. It is a major resource for frozen product, caught by Japanese, Korean and Russian factory vessels. Some product is now available from Alaska, but it is unlikely that much of it would be shipped fresh.

YELLOWTAIL FLOUNDER: *Limanda ferruginea.* This is a right-eyed flatfish from the North Atlantic, of great importance commercially from Rhode Island northwards to Labrador. It is the single most valuable flounder resource of the northwest Atlantic. The largest amount (and generally the largest sizes) of yellowtail are produced in early spring and late fall, but there is a year round fishery.

The yellowtail is one of the prime flounder species. It grows to about three pounds, although most commercial landings are fish between one and two pounds. The New Bedford daily auction establishes the prices for fresh yellowtail. The fish is graded according to the number of dressed fish in a 120 pound box. The largest are generally 90/100 count, the smallest 190/200. Yellowtail flounder gives a firm and sweet fillet which is highly regarded, especially in the midwest and the northeast.

Because of its economic importance, yellowtail is watched carefully by government agencies for signs of over-fishing and there are state and federal limits on catches.

This listing does not, of course, cover all possible flatfish sold as sole or flounder, only the major ones whose names are either legally or generally accepted. Alaska is beginning to develop some major flatfish resources, including some of those mentioned as well as others, such as flathead sole and Alaskan plaice. Some of these species might well figure in fresh supplies in the future, although for the present most of the catch is being frozen because of the logistical problems of getting it to major markets in any other form.

Since fresh imports from other countries are limited, by the cost of freight, to high value products, many alternative flatfish have not been seen on the U.S. market except as frozen product. Some of the flounders from Asia, for example, are poor: the fillets are thin and lack flavor. There are other species, available from Europe as well as from Africa and Asia, which are excellent fish. Some of these are offered from time to time, shipped by air to major markets.

It is worth watching for unusual soles and flounders. Names and customs have little bearing on the actual quality in use of the product. It is (nearly) always worth trying (small) quantities of anything new that may be offered.

FRAUD IN SEAFOODS
Fraudulent dealings occur in every business, from politics downwards (or upwards). Seafood is no exception. Fraud can be separated into financial and product groups.

FINANCIAL FRAUD: Basically, with many variations, this can be defined as buying product with the intent not to pay for some or all of it. The perpetrator sells the product, pockets the money and fails to pay the supplier. In most instances, the buyer claims product imperfections and pays less than the full bill. In other cases, the buyer leaves the business or disappears without paying for any of the product.

This may be brought on by financial difficulties. If the principals of a company have themselves guaranteed corporate lending from a bank and realize their company is likely to go bankrupt, they may be tempted to acquire product, sell it for fast payment and pay off the bank or other secured creditors. This leaves the trade creditors, who are unsecured, with little or no assets when the company is liquidated. It is extremely difficult to prove intent in such cases, which are rarely if ever prosecuted.

Even worse, there have been instances of deliberate scams designed to acquire product, sell it for cash, and then disappear with the proceeds. While proving dishonest intent may be easier in such cases, it is often hard to find the criminals.

To protect yourself from both failing and dishonest customers, check every aspect of credit with great care. Remember that even good customers can get into difficulties and become poor credit risks. Update credit files on existing customers on a regular basis. Using the services of an experienced and specialized credit reporting agency is highly recommended.

PRODUCT FRAUD: The federal government describes this as economic fraud. There are three aspects: substitutions, short weight and inferior quality or size.

Substituting an inferior product, such as greenland turbot for sole, is a complex topic throughout the industry. Because of the enormous number of species available, it is impossible for buyers to be familiar with

FRAUD IN SEAFOODS

all aspects of all of them. Any product which is in strong demand and fetches a high price is vulnerable.

Substitutions of whole fish are the easiest to spot, because the identifying features of head, fins and scales are more or less intact. Nevertheless, many different species of snapper may be shipped instead of genuine red snapper and pomfret may be sold as the more expensive pompano. Headless, dressed fish is harder to identify, although sometimes the fins aid positive classification. Fillets, even with the skin on, can be very difficult to identify. Haddock has a black lateral line, while other gadoids have a white line. However, if the skin is removed, it is impossible to tell haddock from cod or whiting without laboratory testing, using iso-electric focusing or monoclonal antibody techniques.

Shellfish substitutions are also possible. The identifying features of shrimp are normally removed with the heads. Even experts sometimes need to examine shrimp genitals to be sure of the species. Lobster tails, however, can be identified from the tail alone (see Austin Williams, *An Illustrated Guide to Lobsters of the World*).

Throughout this book, possible substitutions are mentioned. Be aware that new ones are often discovered. As relative prices and scarcities change, so do the possibilities for substituting product.

Also bear in mind that some substitutions offer perfectly good product; it is simply not what you expected and paid for. If Chinese white shrimp is good and similar enough to be passed off as domestic, then you will be better off if you simply buy and use the less expensive shrimp. Describe it accurately when you sell it and everyone benefits except the substituter, who has lost your business.

Substitution means the dishonestly passing off an inferior product as a superior one. The terms inferior and superior refer to market perceptions of product as indicated by the price. In other words, a superior product is one that costs more. For example, many people prefer the taste and texture of Alaskan snow crab to king crab, but since snow crab sells for less money than king, no one would ever dream of passing off king crab as snow crab: the market perceives king crab to be a better product, expressing that perception by paying more for it. For the purposes of this definition, a superior product is one which regularly secures a higher market price. Taste and other subjective judgments have nothing whatsoever to do with it.

Substitution only occurs when a less expensive product can be supplied in the place of a more expensive one. If greenland turbot is one dollar per pound cheaper than yellowtail flounder, then turbot may appear labeled as sole or flounder. When the two fish are close in price, there is no motive for anyone to make the switch. Very cheap products will be honestly represented: there is no profit in playing games. It is the most expensive products that are most likely to be the subject of misrepresentation since there is the attraction for the perpetrator to make a substantial profit if he escapes detection and retribution. Although frozen product offers rather more scope for substitution than refrigerated product, there are ample opportunities for supplying misrepresented fresh products.

The first line of defense is to look at everything you receive. This not only builds up experience with assessing products, it also gives you the opportunity to check quality. So long as suppliers know that you look at everything, a dishonest one will be less likely to send you wrong product since the chances of detection are higher. Everything you receive should be checked for weight, grading, quality of material, quality of workmanship, and, of course, for what it actually is. For a detailed, illustrated guide, see this author's *Fish and Seafood Quality Assessment: A Fish and Seafood Quality Assessment: A Guide for Retailers and Restauranteurs.* published by Van Nostrand Reinhold/Osprey Books, 1991.

The second line of defense is knowing what products might be subject to substitutions. These change all the time; it depends entirely on relative prices.

The third line of defence is background knowledge of what is happening in the seafood industry. You should know what fish are in season at any moment; it is surprising how often you may be offered a fish that is not normally available at that time of year.

Finally, if all else fails and a dispute cannot be otherwise resolved, laboratory tests such as iso-electric focusing are available.

Deliberate short weight or short shipment can be very costly. It is particularly easy with fresh fish which is buried in ice, since it takes a lot of work to wash off the ice, weigh the fish and then re-ice it. It is worth checking incoming shipments regularly. See also WEIGHTS AND LABELING.

Inferior quality or size can also be costly. If larger fillets cost more than small ones, check that the size shipped is actually within the range that you ordered. Poor workmanship greatly reduces the value of fresh seafoods. Watch for ratpacking, which is the practice of packing containers with good product on top but poor quality product underneath. It is important to empty out containers and examine the contents. Glancing at the top layer of product will not always suffice.

The major protection against these types of fraud is knowledge. The more you know about fish and shellfish, how it should look, smell, taste and feel and what it should weigh, the better able you will be to spot product deceptions. This topic is examined and illustrated in detail in this author's *Fish and Seafood Quality Assessment: A Guide for Retailers and Restauranteurs,* published by Van Nostrand Reinhold/Osprey Books.

FRESHNESS
Frozen fish is more or less fixed in its quality once it has passed through the freezer. Assuming that packaging and distribution are done properly, frozen product has a long shelf life. Refrigerated

FRESHNESS

product, on the other hand, deteriorates continuously. Even the best fish or shellfish, handled correctly with plenty of ice which is frequently renewed, loses quality and condition as it ages. It is particularly important to be able to determine the freshness of chilled seafoods. The following paragraphs offer some broad guidance.

SMELL: This is positively and absolutely the best way to determine the quality of your fresh seafood. Odors tell you if the product is aging or stale, or if it has been packed poorly. The wrong odors signal a problem.

Before you can use your sense of smell properly, you must train it. There are texts that explain that fresh fish has a pleasant odor, or, more often, no odor at all. This is inaccurate. Most people who write such things have been involved with seafoods for a long time and have grown accustomed to the various smells. Maybe they really even like the smells. Most of us can assess seafood odors without regarding them as pleasant. Poets do not eulogize the smell of the sole and perfume manufacturers ignore the lure of Essence of Catfish. Perhaps "not unpleasant" would be a more acceptable description.

However you regard and describe seafood odors, it is important to train the sense of smell to recognize when seafood is good and when it is not. The FDA teaches people who are called "noses" how to decide when seafood is decomposing. Trained noses can be remarkably consistent in their opinions.

Good fish has, then, a not unpleasant odor. It may be described as seaweedy, which means like fresh, living seaweed, not like heaps of decomposing vegetation on the shore after a storm. The only way to find out is to smell fish at every opportunity. Using common sense and the indicators discussed in this book, you will rapidly learn to tell by your nose the quality of the product.

Off odors are easier to recognize. But you have to learn to distinguish between true "off" smells and the strong, often similar smells that build up inside closed packs and can knock you back when you open them up. Remember that when you first open a container which has been traveling for a while, there will inevitably be a concentration of odors because there has been no opportunity for them to escape. First impressions of the overall shipment, therefore, should be checked by rinsing some pieces and sniffing them again. If the odor is more than a temporary feature on the surface, then the fish is stale.

If you are buying whole or dressed fish, the place where off odor first appears is around the backbone between the gills and the gut cavity. Snap the fish backwards at this point and then smell the exposed meat. There is no particular place to sniff fillets.

Scallops are one of the smelliest seafoods. Even the freshest scallop has a strong (though not particularly unpleasant) odor and in transit this can build up to become quite overpowering. It is essential to get used to the particular odor of scallops in order to be able to recognize when they are turning. Any trace of sourness or iodine in the odor indicates they are stale. Lack of odor on scallops may indicate that remaining shelf life may not be very long.

Whole and Dressed Fish

It is far easier to determine the quality of fish that still has its skin and head than it is to determine the quality of fillets. For those times when you have whole or dressed fish to examine, look at the following features, in addition to what you look for with fillets.

Gills: Normally, gills should be removed before the fish is shipped very far, since spoilage starts there and spreads rapidly due to the amount of oxygen present. If the fish still have gills these should be bright red and clean looking. Pink gills are also acceptable. Later, gills begin to turn gray and then brown. Brown is definitely not acceptable. Gray is marginal and you should evaluate other factors before deciding whether the fish is of adequate quality. Since gills decay so quickly, the condition is not necessarily indicative of the flesh.

Gut Cavity: To determine the condition of the fish, look for clean and rather glossy surfaces inside the gut cavity. If the belly bones are torn from the flesh or can be torn away easily, the fish may be belly burned. Fish in good condition will have the belly wall bones adhering tightly to the flesh.

The workmanship of your processor is also revealed by the gut cavity. All the guts should have been removed. Heads should be cut off cleanly without wasting edible flesh. The cavity should be washed and clean. There should be no cuts in the flesh of the fish.

Slime: Fish naturally have a protective coating of slime which protects them in the water and helps to oil their movements. This is why a fish is so slippery when it is first caught. Generally, the slime will have been washed off long before a fish reaches a distributor. If slime is still on the skin, it should be transparent, like water. Slightly milky slime is also acceptable. Yellow or brown slime is not acceptable.

Skin: If slime is present you can also evaluate the skin, which should be shiny and bright. Dull skin is marginal. Bleached skin and skin which looks as if it is a little too large for the fish are bad signs. However, if the fish has been washed and scaled, as most of it is before shipment from the port, the skin will NOT be bright and shiny, so you must be careful how you read this indicator.

Eyes: If the eyes are still around to look at, they should be convex with translucent corneas and they should shine a little. Flat eyes, becoming opaque, indicate marginal quality. Sunken and discolored eyes call for rejection of the fish.

Temperature: Correct temperature and plenty of ice are vital to maintaining the quality of fish. Check the temperature of the product when it arrives. Use a probe thermometer inserted into an unfilleted fish through the back as near to the backbone as possible. Test the thickest part of the fish. Fillets should be checked in a similar manner. Insert the probe into the center of a thick fillet. Make sure the bulb is completely covered.

The temperature of the fish, whatever the season of the year or the temperature outside, should be 32°F.

There should be plenty of ice. In hot weather, equal weights of ice and fish may be required. Even in cold winters, the minimum amount of ice should be one pound of ice for three pounds of fish. Replace ice when fish arrives. The old ice will be building up bacteria from the fluids dripped from the fish and the bacteria will speed decomposition. Investment in sufficient ice is the most prudent action of a fresh fish user.

There are less subjective ways to measure freshness, although so far all the methods have limitations. There are chemical tests for measuring hypoxanthine and other products of decomposition. There are electrical devices which measure changes in the fish flesh as it decomposes. The problem is that such tests have to be tried on a huge variety of fish in order to produce consistent and verifiable results. A device or test may work on several species but not on many others. Until more work is completed on these technologies, subjective tests of smell and appearance are likely to remain the most successful way to assess the quality of fish and shellfish.

Detailed indicators of freshness for different products are mentioned under appropriate individual headings throughout this book.

FRIGATE MACKEREL
Auxis thazard. Frigate mackerel is a different fish from bonito. However, it looks and tastes very much the same as bonito. For commercial purposes, it is effectively the same fish.

FROGS
Few fresh frogs are produced in the United States, although Louisiana used to be an important supplier. Resources are available from central Florida, for example, but the cost of domestic frog legs is very high. It is possible that aquaculture techniques will restore fresh frog supplies. Until that time, the market will continue to depend on frozen legs from Asian countries.

G

GAPING Separation or partial separation of the individual flakes of a fillet, so that slits or holes appear. In bad cases, fillets may fall apart completely if the skin is removed. Fish muscle is composed of flakes. Each flake is separated from the next one by a thin, shiny membrane. Connective tissues run through each flake to these membranes. If these connecting threads are strained, they may break and then the fillet gapes.

Gaping affects the usability of the fish. Haddock, cod, whiting and pollock, which all have quite large flakes, are among frequent victims of gaping. Fish with large flakes and soft flesh, such as bluefish, are especially prone to gaping. Flounder, sole and ocean catfish suffer very little. Gaping affects the appearance of the fillet and reduces its marketability. It also affects how usable the fillet is: pieces which fall apart cause obvious problems for anyone trying to cook them.

Gaping is caused by a number of different factors, including the following:

- Fish frozen whole and then thawed and filleted has a tendency to gape more than fillets produced from fish that has not been frozen.
- Rough handling of fish, including standing on them or throwing them around, causes gaping.
- Rigor mortis—the stiffening of the body after death—begins shortly after the fish is landed. The onset of rigor mortis and the time it lasts are both extremely variable. Fillets cut from a fish before it has passed completely through rigor are quite likely to gape. This is because the stiffening process needs the skeleton to hold the muscle structure together. If the fillet is removed, the rigor effects pull apart the flakes of the meat.
- Fish that are not chilled quickly after catching and kept cold will always gape much more than fish that have been properly handled.

The connective tissue is damaged by higher temperatures as well as by the rapid onset of rigor mortis.
- For some reason, small fish tend to gape more than large ones of the same species in the same condition. Possibly the connective tissue is relatively stronger in larger fish.
- Fish that have recently spawned have soft, watery flesh. When they resume feeding, and the flesh quality is beginning to improve, gaping is most likely to be a problem. Fish in this condition must be handled with particular care.

Most gaping is caused by errors in handling the fish. Gaping fillets should be carefully examined and, if necessary, rejected.

GEFILTE FISH Fish balls, a traditional and popular item in Jewish cooking. Pike and whitefish are very frequently used, but there are numerous recipes, mostly based on freshwater fish.

GIBBED Small fish such as herring are sometimes processed this way: the gills and viscera are removed through the gill flap, leaving the head on the fish and the ore or milt inside it.

GOATFISH *Mullidae*. This is the fish called mullet, used and highly prized in Mediterranean countries. It is different from the various Florida region mullets, which are true mullets (*Mugilidae*), not goatfish. Goatfish are excellent small panfish with few bones and firm, white flesh, but although they can be caught along most of the Atlantic coast and in southern California, they are seldom eaten in the United States, except in Hawaii. Small quantities are sold on the New York market.

GONADS Sexual organs, of fish as of all other animals. The (female) roe of many fish is eaten: caviar is of course the prime example, but salmon, cod, herring, shad, alewife and many other fish supply excellent roe. Milts, the male equivalent, sometimes called soft roes, are seldom eaten in the United States, but herring milts are popular in Canada as well as in Europe. Sea urchin gonads are the only part of the animal that is eaten at all.

GOOD MANUFACTURING PRACTICE GMP is a basic outline of how to set up and run a food processing facility. GMP covers personnel, buildings and equipment, production, warehousing and distribution. GMP is printed in Title 21 of the Code of Federal Regulations, Part 110. Government purchasing and other standards for many food products refer to GMP as part of the specification.

GRAS GRAS stands for Generally Recognized As Safe. It is a list of substances, including additives, which are acceptable for use in foods. Salt and pepper are on the GRAS list. Substances not on the list may be used only if they have been specifically approved by the FDA. Even though a substance is listed as GRAS, it still has to be used properly and in accordance with accepted and acceptable practices.

GRAS substances include flavorings, nutrient supplements such as vitamins and minerals, anti-caking agents, preservatives, emulsifiers, sequestrants and stabilizers. The list is extensive and changes occasionally. The current list is available in the Federal Code, Title 21, Part 182.

GRAVLAX Also called gravadlax, dill salmon, marinated salmon. Fillets of salmon marinated in a dry mixture of salt, sugar, white pepper and dill weed. Similar to the finest cold smoked salmon in texture and taste, gravlax originated in Scandinavia as a traditional method of preserving salmon. It is finding increasing acceptability in the United States as a gourmet item.

Gravlax is readily available in both institutional and retail packs. It can also be made quite easily at the restaurant level. It is important to use only very fresh salmon, in good condition. Atlantic salmon and Pacific kings both make first class gravlax.

GRENADIER *Macrouridae.* Grenadier is sometimes called rattail, but not by marketers hoping to sell it. There are some 260 species found in very deep water throughout the world. Most are small, the largest growing to less than three feet in length. There is thought to be a very large resource of grenadiers off the California coast.

GRILSE Atlantic salmon returning to their home stream for the first time. (Although Atlantic salmon return more than once to spawn, all the Pacific species normally return one time only, dying after spawning.) Grilse are younger, smaller fish and every bit as palatable as the older and larger ones. Because they are smaller and less oily, they are less desirable for smoking and may therefore be a little cheaper. See SALMON.

GROUPER Particular species of grouper may also be called gag, jewfish, sea perch, scamp, hind and warsaw. Groupers are in the sea bass family and overlap sea bass in nomenclature. Sea bass names approved by the FDA are shown in the *Fish List* table in the Appendix. Most groupers are caught from the Carolinas, around Florida and through the Gulf. Groupers are reef fishes. Yellowfin and misty groupers may cause ciguatera poisoning, but other groupers appear safe. This problem is in any case largely restricted to the West Indies and the

GROUPER

Bahamas. Groupers from places as far away as Oman are imported fresh by air, so the range of origins and species that you might be offered is immense.

Groupers generally have white, very firm flesh and are excellent fish to eat. There are no small bones, but the skin should always be removed before the fish is cooked as it is strongly flavored and tough. Fresh supplies are available dressed or filleted. Large fillets may be cut into smaller, more convenient sizes. Some species of grouper grow to many hundreds of pounds; large fillets are not rare. Grouper is landed throughout the year but is most plentiful in the summer months and least plentiful in January and February.

Commercially marketed fish (all of which may be called grouper in addition to the names given here) include:

BLACK GROUPER: *Mycteroperca bonaci*. Fairly common in southern Florida, the black grouper is usually about 20 pounds when marketed. Because it comes from rather deep water, it tends to have fewer parasites than some other groupers.

GAG: *Mycteroperca microlepis*. Similar to the black grouper, and sometimes called black grouper, it may be distinguished most easily by the shape of the tail, which in the black grouper is straight at the end. The gag has a tail with a curved, concave end.

JEWFISH: *Epinephelus itajara*. Also called spotted grouper, giant sea bass. The largest grouper, sometimes reaching 700 pounds (although 50 pound fish are more common). The flesh is particularly good. It is found mainly in southern Florida and the Caribbean. The warsaw grouper (see below) is similar and the two are often interchangeable on commercial markets.

MISTY GROUPER: *Epinephelus mystacinus*. A large, deep water species from the Atlantic and southern Florida, this is not particularly common.

NASSAU GROUPER: *Epinephelus striatus*. Fairly common, especially around the Bahamas (which explains the name). Generally marketed around 10 pounds, although the fish grows very much larger.

RED GROUPER: *Epinephelus morio*. Fairly common in southern Florida and the Gulf, red groupers reach 50 pounds and are an important commercial fish.

RED HIND OR HIND: *Epinephelus guttatus*. A small grouper, usually between two and four pounds, the red hind (or strawberry grouper in the Bahamas) is an excellent eating fish.

ROCK HIND OR HIND: *Epinephelus adscensionis*. A small grouper, found around Florida and particularly around the Bahamas and in the Caribbean.

SCAMP: *Mycteroperca phenax*. A small, commercially significant grouper, mainly from Florida and the Gulf coast.

SNOWY GROUPER: *Epinephelus niveatus*. The species is sometimes called golden grouper, though this name is not approved by the FDA. It is a large and deep living fish found in southern Florida and through the tropical parts of the Atlantic.

SPECKLED HIND OR HIND: *Epinephelus drummondhayi*. Similar to the red hind, the speckled hind is landed and generally available in the Carolinas and in the eastern parts of the Gulf.

WARSAW: *Epinephelus nigritus*. Also called black jewfish; the term is not approved. The species is similar to the jewfish, though seldom exceeding 300 pounds. Warsaw and jewfish are virtually interchangeable commercially.

YELLOWEDGE GROUPER: *Epinephelus flavolimbatus*. A medium size grouper from Florida and the Caribbean. As the name indicates, it has yellow edges to the fils.

YELLOWFIN GROUPER: *Mycteroperca venensoa*. Similar to the yellowedge.

YELLOWMOUTH GROUPER: *Mycteroperca interstitialis*. A common, small grouper in Florida, most fish are around two pounds and seldom exceed four pounds. It is good quality meat.

Strawberry grouper and Kitty Mitchell are other grouper names commonly marketed and there are other names, too. Some of these are regional names for the species described above. There is little market distinction between any of the groupers and mainly small differences in taste and texture, so the precise species identification is, perhaps, not too important. Reddish groupers are preferred in some areas, blackish ones in others. Any differences are unlikely to be more than a few cents per pound.

GURRY
Waste material from processing fish. It may include heads, bones, skin and guts. Sometimes used for feeding minks or for lobster bait. It is turned into fishmeal if the local volume, as well as the neighbor's sentiment, is sufficiently strong to support a fishmeal plant. Otherwise, gurry presents major problems for disposal in an acceptable manner. Work to utilize gurry proceeds. The most promising direction may be using a simple silage process to turn it into concentrated fertilizer for domestic and horticultural use.

H

HADDOCK *Melanogrammus aeglefinus.* Haddock is a close and very similar relative of the cod. Haddock are generally smaller than cod. Few fish exceed 20 pounds and most now are under five pounds. The flesh may be a little softer than cod, but it is virtually impossible to tell them apart when they are cooked. Haddock has a distinctive black thumb print on the kin on each side and a black lateral line. Cod, pollock, and whiting have white lateral lines. Because the skin is really the only way to tell haddock apart from cod without laboratory analysis, processors will tend to leave the skin on haddock fillets to prove that is what they are. Skinless haddock is frequently offered, of course, but is probably shipped less frequently than it is invoiced. Normally, if the word can be used about any market for fish, haddock is a little more expensive than cod, though that is by no means a firm rule and there are many occasions when cod is more expensive than haddock.

Haddock is a North Atlantic fish, with a large proportion of U.S. fresh and frozen supplies coming from Canada. There is no equivalent Pacific species. Fresh fillets are imported from Iceland and Norway as well as from Canada.

The Boston fish pier has established size descriptions for dressed haddock. Most other ports use these definitions also:

Snapper haddock is under 1½ pounds, dressed.
Scrod haddock is 1½ to 2½ pounds.
Large haddock is 2½ pounds and over.

The term jumbo may be applied to particularly big haddock. As with all such definitions, it is preferable to specify a size by weight rather than by a description as this ensures no misunderstanding.

Haddock is landed all year, with peak supplies in May and June and generally large supplies during the spring, summer and early fall. The resource has been under heavy pressure for many years, but there is some sort of cyclical pattern to catches.

Halibut
Hippoglossus hippoglossus (Atlantic halibut), *H. stenolepsis* (Pacific halibut) and *Paralichthys californicus* (California halibut). Greenland halibut, black halibut and any other descriptions are misrepresentations of fish that are not halibut but greenland turbot. California halibut is from a different family of flatfish, but tends to sell exactly as halibut.

Atlantic halibut has virtually disappeared from U.S. waters, although some is still caught in Canada and there are occasional fish landed in the United States. At one time, huge fish weighing up to 700 pounds were caught regularly in northeast areas. The small quantities of small fish that are caught now are sold fresh into local markets, headless and dressed for filleting and steaking. The meat is excellent: firm, white and well flavored. Atlantic halibut is one of the very best of all seafoods. Much of the fish caught now is incidental catches by fishermen looking for other species.

Atlantic halibut are generally offered as graded fish, although because of the small quantities this is simply a way of describing what is available rather than of offering the buyer a choice. The color of the underside skin varies between white and gray. White fish are preferred, although the flesh is identical. Larger fish yield more edible portions than smaller ones.

Pacific halibut is equally good. Caught from Oregon to northern Alaska by longlines, this resource is also under heavy pressure. International and national rules restrict fishing efforts severely, sometimes to only a few days each year. The United States has adopted a peculiar system for limiting halibut catches. Boats are allowed to fish during a short season for a limited amount of fish. This encourages large numbers of boats to enter the fishery to catch as many fish as they can before either the quota is reached or the time limit expires. The result has been that there are far too many boats involved in the fishery, which increases prices to buyers. Another result has been that the fishery is especially dangerous; the limits encourage fisherman to stay at sea in bad weather and every year boats and crew are lost.

There are moves to change the system and control halibut landings in ways which would permit landings of the same amount of fish over a much longer period. The present method effectively precludes a market for fresh halibut, since supplies are available for only a few days or weeks every year. British Columbia fishermen catch halibut over a much longer period. Small quantities of fresh halibut are available from that source, but the Canadian share of the total fishery is not large.

Pressure on halibut resources increased with the development of large freezer trawlers targeting Alaska pollock. Juvenile halibut were caught with the pollock, making management of the halibut fishery particularly difficult. Gear technologists are working on methods to catch pollock without also catching young halibut, a development which would help both fisheries.

Halibut is landed mainly in Kodiak, southeast Alaska, British Columbia and around Seattle. May and June is the peak season, depending on when the fishery opens. British Columbia and the United States take alternate years to open in May or June. Pacific halibut of up to 500 pounds have been caught, but usually fish are between ten and 100 pounds.

Halibut sold fresh is generally headless and dressed. Fletches may be offered: these are fillets, though, of course, much larger than the fillets from other fish. Steaks are cut either from the dressed fish or from fletches. Yields from larger fish are better, but buyers are also reluctant to handle very heavy halibut unless they have suitable equipment. It is necessary to trade off yield against the handling problems.

California halibut seldom exceeds 50 pounds. The meat is good, but inferior to that of Pacific halibut. Most of this fish, which technically is a flounder, is sold as fletches.

HANDLING FRESH SEAFOODS

Proper handling of fresh seafoods is at least as vital to profitability as any other part of the seafood business. A great deal of money is lost when product goes bad before it can be sold. Good handling (and packaging) are essential to maximize shelf life.

The simple rules for handling fresh fish and shellfish are: COOL, QUICK AND CLEAN. Keep it COOL. Handle it QUICKly. Keep it CLEAN.

RULE 1: COOL: Ice and temperature. Proper icing has the most to do with retaining fish quality. Refrigeration on its own is not sufficient. If you put fish in the cooler, the fish will rapidly dry up, then curl and finally turn yellow and brown. This is unattractive and unprofitable. The cooler firstly is not cool enough and secondly is not damp enough. Fish need the wet coldness of ice if the flesh is to be kept in peak condition.

This does not mean that you can just put a layer of ice on top of the fish and leave it at that. The ice and the fish have to be thoroughly mixed together. A box of fish should have ice on the bottom, ice mixed throughout the fish and a thick layer of ice on top. This will hold the temperature of the fish at very close to freezing point (32°F), which is where it should be.

You need to use at least one pound of ice for three pounds of fish in the winter; and in hot weather, as much ice as fish. In between, two pounds of ice for three pounds of fish should be sufficient. You cannot spoil fish with too much ice. If in doubt, use more.

The type of ice you use is also important. Flake ice is good because it

HANDLING FRESH SEAFOODS

has a great deal of surface area to melt and comes into contact with the product better than other forms. Large blocks of ice, chipped by hand or machine into irregular lumps, are not very good. The sharp edges of some of the pieces can damage the fish and the larger lumps prevent good, even contact between the ice and the whole surface of the fish. Ice cubes and tube ice designed for dry martinis should also be avoided. In an emergency, any ice is better than no ice, but you should be equipped to make or buy the ice you need.

Ice has to be changed from time to time, costly as this seems. Blood and other fluids dripping from the fish provide breeding grounds for bacteria which hasten spoilage of the fish. Dirty, stained ice is a problem and one that is easily solved by throwing out the bad ice and replacing it with good ice. There are no firm rules for how often to do this; common sense should tell you. If it looks nasty, it certainly is.

Fillets must be placed in plastic or metal containers and not mixed directly with ice. Direct contact with ice leaches proteins and spoils the texture and appearance of the fillet. Metal containers are best because the material allows the fastest heat transfer away from the fillets. As the ice melts, it leaves a cavity between the fish or the containers of fish and the ice as it melts away from the fish. This creates an air pocket between the fish and the ice, which means that the temperature of the fish will increase. Only good physical contact between the fish and the ice will ensure that the temperature is kept at the required freezing point. To overcome this problem, the ice again must be changed or at least replenished frequently. Replenished does not mean just putting more ice on top. It means removing the fish and re-icing the whole box. Merely adding ice to the top will leave some of the air spaces between fish and ice. It is these air spaces that you are trying to remove.

There are certain antibiotics than can be added to ice to reduce bacterial growth. These are illegal in the United States and may also be dangerous to some of your customers.

It should hardly be necessary to state that fresh fish and shellfish should be kept refrigerated at all times. Ice is vital, but refrigeration preserves both the ice and the fish. While ice without refrigeration will look after your product, the higher melting rate will "wash" some of the weight (in the form of soluble proteins) out of the fish. And, of course, you will need to replenish and replace the ice sooner. If the iced product is kept in the cooler, the ice melts more slowly. This is better for the product and uses less ice.

All holding areas and trucks should be refrigerated. Processing and working areas should be kept cool.

Fresh and iced product should never be put in a freezer, or in a truck maintained at temperatures suitable for frozen product. To freeze product, you must have equipment capable of taking down the temperature very quickly, which is not the case in any storage freezer. If you put fresh

HANDLING LIVE MOLLUSCS

fish in the storage freezer it will freeze slowly, which spoils its texture. If the product is thick, such as a dressed or steaked fish, it may freeze so slowly in the center that the flesh may actually have time to turn bad before it is hard frozen throughout, even though the outside will appear hard and cold.

Iced fish left in a freezer for storage purposes builds a solid crust across the top of the container, which prevents proper air circulation and which also damages any product it contacts.

RULE 2: QUICK: You have to sell fresh product as soon as you can. Modified atmosphere packaging, used successfully elsewhere to prolong shelf life, is still not permitted in the United States. Speed is important in the physical handling of seafoods from place to place. Move it into the cooler as soon as it arrives. Waiting until someone has time to re-ice it may mean that it warms up enough to deteriorate significantly. A difference of 8°F can lose several days' shelf life. The same applies to loading trucks. Never assemble orders on the loading dock, or in a warm place. Keep everything in the cooler until it is needed, then move it quickly to the next place, whether that is the truck, the showcase or the kitchen.

Ensure that product is dated when it arrives, and that you ALWAYS use the oldest first. This means keeping careful records of arrivals and shipments. These records will not only help to ensure that fish does not go bad because it is accidentally forgotten at the back of the cooler, they will also help to reduce pilfering since shortages will come to light very quickly. Both benefits improve the bottom line.

RULE 3: CLEAN: The key to selling fresh seafood is to move it fast, while it is still good. Keeping it clean will give you extra time to move it, because spoilage is much faster if the product is dirty and contaminated. For a properly clean operation, buildings, equipment and people must be right. But even the best designed premises will become dirty unless regular procedures are followed to ensure sanitation.

HANDLING LIVE MOLLUSCS
Clams, mussels, oysters and periwinkles are all molluscs and are commonly sold live, in the shell. They are either eaten raw or are cooked from the live state. Some of these animals survive out of their natural seawater habitat for long periods if properly handled. Shipment live is still probably the best way to distribute them, although if frozen quickly by modern methods they can be just as good as live shellfish, and easier to handle. Users and consumers have strong prejudices against whole frozen shellfish, since it is impossible to demonstrate that the animal was alive when frozen. Even when that problem is resolved, there will remain a market for the live shellfish, so it will still be necessary to know how to handle it.

Oysters are probably the most delicate of the shellfish that are commonly handled live. Mussels are also quite delicate and are most vulnerable to rough handling, because of the thin shells. Since they are much cheaper than oysters, they are more likely to be carelessly mistreated. Soft shell clams are almost as delicate as oysters, while hard shell clams are much tougher, especially the very big ones, which are definitely hardy. The toughest of all is the snail (scungili), which seems to be able to live through almost anything. The meat can be pretty tough, too.

Temperatures should be maintained below 40°F and preferably at 35°F. (Note that live shellfish harvested from warm water may need to be kept at warmer temperatures if they are to survive, even though this may cause increased bacteria growth.) Live shellfish are particularly vulnerable to dehydration, which kills them. Use plenty of ice to keep them cool, but remember that freshwater also kills them. It is essential to keep live shellfish away from pools of melting ice. Store containers on pallets so that the melt water drains away from them.

Ice use in retail displays should not be allowed to drip or melt on to any live shellfish, since fresh water may kill them. Shellfish should be laid on top of the ice in a display. Since drip soon makes the ice unattractive, it is better to put the shellfish on a metal tray which conducts heat well and put the tray on the ice. Never throw ice on top of the display at night to preserve it until the next day. The few minutes it takes to put the shellfish safely away in the cooler, properly protected in containers, is worthwhile in terms of preserving product suitable for sale.

Handle bags of shellfish as though they were eggs. Shocks are bad for the animals. Dropping the bag even a few inches can crack shells at the bottom of the bag. Shellfish with damaged shells do not live long and should be rejected by the consumer or the chef. Shells may also be smashed, and the fish killed, by quite moderate bangs and knocks. Shocks that do not damage shells still appear to shorten the life of shellfish.

Some molluscs will live for lengthy periods in tanks of circulating sea water. These are technically subject to the rules of the NSSP. Never discharge used water from a tank into the sea. Live tanks are popular displays in retail stores as well as restaurants. However, they must be properly managed and maintained. One state (Texas) has already banned sale of shellfish from live tanks and others may follow suit. The problem is that if tanks are not run well, bacteria in the shellfish can multiply to dangerous levels.

For more information on handling molluscs, see entries for individual shellfish and the one on the NATIONAL SHELLFISH SANITATION PROGRAM.

HANDLING LIVE CRUSTACEANS
Lobsters are normally shipped and sold alive. For information on handling live lobsters, see LOBSTER. Dungeness, blue and jonah crabs may also be kept alive for sale. For information, see headings LOBSTER and CRAB.

H & G   Abbreviation for headless and gutted. H & D, or headless and dressed, means the same thing.

HATCHERY FISH   The term is commonly applied to Pacific salmon which have reached the spawning grounds, released their spawn and are dead or dying. The fish are spent and have used most of their reserves in swimming up the river, during which time they do not eat. The flesh is soft, colorless and not really palatable. Hatchery salmon may be used for animal food or for manufacturing pastes. It is not suitable for regular human consumption.

HERRING   *Clupea harengus harengus* (Atlantic herring). *Clupea harengus pallasii* (Pacific herring). The Atlantic and Pacific herrings are subspecies, which means that they are virtually the same fish.

Other names and other fish: River herring is alewife. Sardine, pilchard and anchovy are all confused with herring, and from time to time, fish labeled with these names may be herring. Juvenile herring are canned on a large scale and marketed as sardines in Maine and in Canada.

Herring was once a cheap staple food in northern Europe, where huge quantities were pickled, smoked or otherwise preserved for use throughout the year. Because of the dramatic collapse of the European resource, herring is no longer cheap and many variations of the traditional cured products now qualify as gourmet and luxury items.

On the west coast, herring is caught off Alaska, British Columbia and Washington for export to Japan, where the roe is highly prized. Quantities of Atlantic herring are also used in this way, but the eggs are smaller and thought less desirable for that reason. Roes and milts of Atlantic herring are frozen separately by processors in Atlantic Canada and sold in Europe, where both items are favored products. The most expensive roe product, limited to the west coast, is roe attached to kelp. The herrings spawn their eggs onto kelp fronds. These are harvested together with the eggs and are sold as *kazunoko-konbu* to Japanese markets in the United States and Japan at very high prices.

In the Atlantic, Canada catches large quantities of herring, most of which is frozen and exported to Europe. The resource extends as far south as the Carolinas, but most of the fishing is done in northern areas. The major U.S. ports for herring are Gloucester in Massachusetts and Eastport, Rockport and Rockland in Maine. Herring fishing is highly seasonal, although the development of large fishing vessels using midwater trawls has greatly extended the season and increased pressure on the stocks.

Herring spawn in late fall. As the fish develops roe (female) or milt (male), the oil content of the fish flesh drops. Since a great amount of herring is used for smoking and curing where a high fat content is essential, fish that are about to spawn are worth much less than fatter fish. Of course, the ripe female fish are normally saleable to Japan for the roe. It is possible to distinguish the male from the female herring and ship only the females.

Mature herring weigh about one to 1½ pounds. Similar fish are found throughout the world, but the North Atlantic herring is probably the most prized because of its very high oil content: as much as 20 percent in some seasons, compared with as little as 3 percent in thread herring from the Pacific coast of South America. Bear in mind that most processors of herring items need a fat content of at least 10 percent and that even this is too low to make top quality products in many cases.

Fresh herring markets in the United States are extremely limited. Partly because of the high oil content, the fish spoils rapidly, as the fat oxidizes (turns rancid). However, newly caught herring in top condition is excellent to eat. It is unlikely ever to be a popular fish in the United States because of the numerous small bones and pronounced flavor.

Herring should be undamaged: torn flesh and bruises, and especially tears along the belly walls, are unacceptable. Although fat content has to be measured by laboratory techniques, the presence of well developed roe or milt in the fish is a sure indicator that the oil content is comparatively low. Just after spawning, the oil content is at its lowest. Such fish can be recognized by their lackluster and emaciated appearance.

At the period when the fish is fat, it is also feeding heavily. Undigested food in the stomach will ferment and form gas, of which the early signs are distended bellies. Fat herring that is intended for fresh use should preferably be gutted and well iced as soon as possible. Mostly, herring is shipped whole, including viscera and heads.

Smoked herring products include kippers, bloaters, smokies and blind robins. Any of these can be handled fresh or frozen. For further details, see the heading SMOKED FISH.

Pickled herring products popular in the United States, include rollmops, Bismark herring, dill herring and herring in sour cream and wine sauces. There are many more such preparations. Most of these sell through delicatessen outlets rather than through the fish trade. None of them are inexpensive. For both retail and institutional use, many are available in gallon plastic containers. Under refrigeration, pickled herring products will keep for a long time. The best are made from very fat fish. If the product's texture is a little dry and hard, the herring was probably too lean for the cure. Yellowing on the cut surfaces of the herring indicates rancidity.

HISTAMINES
Chemicals produced by decomposition of flesh in certain species in certain conditions. Histamine poisoning is extremely unpleasant and can be fatal to heart patients and asthma sufferers.

HISTAMINES

Because it is sometimes associated with scombroid fishes (tunas and mackerels), it is also known as scombroid poisoning.

Imported mahimahi has been one of the more frequent causes of histamine poisoning. The FDA carefully monitors shipments, but because the occurrence is irregular, sampling can miss affected fish in a batch. Histamines are a product of poor handling. They develop in fish left in warm conditions. The problem can be avoided if fishermen and processors handle vulnerable species properly. If you buy mahimahi and tuna from reliable sources that look after the fish, you should not have a problem.

I

ICE Ice is the most important ingredient in maintaining the freshness and quality of chilled seafoods. There are several different types of ice, which are discussed above under HANDLING. Quantities of ice and how to use it are discussed under the same heading.

INCONNU *Stenodus leucichthys*. The FDA permits you to call this fish trout (it is a salmonid), but disapproves of sheefish. The largest member of the freshwater whitefish family, inconnu are imported from Canada, where they are caught in lakes during the winter. There is also an anadromous strain. Inconnu in lakes grow as large as 20 pounds and the sea-going variety is even larger. The fish is excellent but little known in the United States.

INK Squid, octopus and cuttlefish (inkfish) expel a black fluid into the water to enable them to escape from predators which cannot see them through the cloud of ink. Squid and cuttlefish are both sometimes cooked in their own ink, which demands careful removal of the ink sac, without breaking it. Ink is available frozen in pint containers.

INSPECTION Mandatory inspection of fish and shellfish is an important political and practical issue for the industry. If it comes, it is likely that it will be based on a control system known as HACCP, for Hazard Analysis of Critical Control Points. A HACCP system monitors points in the production and distribution process which may permit contaminants to reach the product. It is a quality control technique which can be closely monitored for effectiveness. HACCP is different from the techniques used in meat and poultry inspection; these are based on sampling of output. Many seafood processors are aware of the need to satisfy customers that their products are safe and wholesome and are beginning to install voluntary HACCP systems to provide that assurance. A joint initiative by the FDA and the NMFS intends to offer a voluntary HACCP-based inspection service to the seafood industry.

Voluntary inspection of seafood plants and seafood products has been available for many years from the U.S. Department of Commerce's NMFS in return for fees. U.S.D.C. inspects a substantial percentage of all the seafood sold in the United States, especially branded products for retail sale and for mass feeding. U.S.D.C. will award inspected seafood a grade of A, B or C, according to detailed scoring schedules based on standard specifications (see Title 50 of the U.S. Code for the full documentation). Grade A means top or best quality and many packers are able to label their output with the Grade A shield. For products where there is no detailed specification, the PUFI mark is available (PUFI stands for Packed Under Federal Inspection).

IRRADIATION

Preservation of food by killing spoilage bacteria with low doses of gamma radiation. Picowaving and ionizing irradiation are alternative terms. Without the spoilage bacteria, food keeps longer; iced fish may keep two to three times longer after irradiation. Irradiation has also been suggested for killing bacteria in live shellfish.

The FDA so far considers irradiation as a food additive and insists that it must be listed on the label of any food so treated. Approval for its use in the United States is limited mainly to spices and to killing trichina parasites in pork. Irradiation industry pressures for greater approval have been intense. Retailers, conscious that consumers are nervous about irradiation, have not pushed for early approval. Consumer groups tend to oppose irradiation and support the idea that its use must be noted on the label. Irradiation is a sensitive political topic; one particularly sensitive problem is how to dispose safely of the spent nuclear material from irradiation plants.

It appears that irradiation produces substances called unique radiolytic products (URPs) in treated foods. These have been neither identified nor tested but are said by irradiation proponents to be safe. For a fuller discussion of irradiation, see *The New Frozen Seafood Handbook*.

Irradiation has been used for a number of years in Holland and Denmark to sanitize seafood products, particularly shrimp and frog legs. Product treated in this way is not allowed to be imported into the United States, although over the years there is no doubt that considerable quantities have entered. One of the concerns is that the unnaturally clean surfaces of irradiated product are prime breeding grounds for new bacteria contamination and that without competition from spoilage bacteria, organisms much more harmful to consumers might flourish. Spoilage bacteria play an important role in warning us that food is not fit to eat. Removing that warning system may not be wise.

ISO-ELECTRIC FOCUSING IEF or electrophoresis is a technique for identifying species by analyzing the pattern of proteins in the flesh. These proteins, when electrically charged, form a pattern that is unique to a species. The technique is approved for raw seafoods by the Association of Official Analytical Chemists. This means that the tests can be considered reliable for use in lawsuits. If you have a major dispute over the identification of a product, IEF tests may be the best way to resolve the matter. The technique is not yet able to distinguish the species of cooked foods.

Monoclonal antibodies can also be used for making positive identifications of fish and shellfish. This is a newer technique.

J

JACKS *Carangidae.* These are silvery, compressed fish, usually pelagic. There are over 50 species called jack by the FDA. Amberjack, jack mackerel, pompano, permit and pomfret are jacks. These are covered in separate entries in this book.

JACK CREVALLE: *Caranx hippos.* An oily fish available in vast quantities in the summer months from Florida. Jack crevalle is landed year round, with the largest production in April and September. It is found throughout the warmer subtropical and tropical waters of the Atlantic: much more is caught and used in Africa than in American countries.

Blue runners and jacks are similar but different fish. Both are better eating when small, generally used as pan fish. They also have whiter flesh when small. Jack crevalle reach over 20 pounds and are a popular sport fish, but both jacks and blue runners are better when they are under two to three pounds. The smallest jacks, around four ounces, are known as cluckers. Flavor is improved if the fish is bled before cooking. This may be done at the plant, by cutting off the tail before further processing, or by the user cutting off the tail and soaking the fish in iced water overnight. Market area is limited to the Gulf region for fresh consumption. Supermarkets offer dressed, pan ready fish. Frozen whole blue runners are also used by zoos and aquaria.

Larger fish are filleted. If the skin and red meat (the oily part) is removed before the fish is cooked, the taste is excellent. Jacks make good smoked product and are sometimes used for Japanese raw fish preparations.

The resource appears to be very large and is not heavily fished.

BLUE RUNNER: *Caranx crysos.* The Pacific green jack (*Caranx caballus*) is very similar—almost identical for commercial purposes. Blue runners are caught in large quantities in Florida year round, with half of the annual landings usually in May. Most fish weigh about two pounds when caught. A fish weighing four pounds is unusually large. It is cheap, normally sold whole, dressed or pan ready. Zoos buy whole blue runners for feeding to their animals and aquarium fish.

The flesh is oily and fairly dark, with a reddish strip along the fillet which contains a lot of fat. If tails are cut off to bleed the fish as soon as they are caught, the flesh will be lighter and better flavored. Removing the red portion from the fillet also improves the flavor. Such processing is seldom feasible if large quantities are being landed on a fishing boat at one time. Smaller fish have whiter flesh. Properly handled and processed, blue runners provide an acceptable and inexpensive fillet or pan fish which deserves wider markets.

Resources of blue runner are thought to be very substantial.

JUMBO A word popular with seafood packers wishing to make their product sound big. Jumbo cod used to be fresh cod of 25 pounds or more and jumbo cod fillets were cut from this size fish (at 40 percent yield, each fillet would weigh at least five pounds). Today, cod fillets as small as one pound may be misdescribed as jumbo cod. Jumbo lump crab meat means the two large pieces from the backfin (body) of the blue crab, which is a rather small crab. Jumbo shrimp used to mean 21/25 count shrimp, but abuse of the term led the NMFS to cancel the use of descriptive size terms for shrimp some years ago. Shrimp sizes should always be defined by numerical counts. Descriptive terms are not only misleading but illegal.

K

KILO One kilogram (abbreviated kg) is 1,000 grams. The American equivalent is 2.2046 pounds. Two kilo packs contain 12 percent less product than five pound packs.

KINGFISH *Menticirrhus spp.* Southern kingfish is *Menticirrhus americanus*, northern kingfish is *Menticirrhus saxatilis* and Gulf kingfish is *Menticirrhus littoralis*. These fish are sometimes, confusingly and incorrectly, called whiting and king whiting. There are many other local names. These fish are found from Maine southward throughout the Gulf. Most commercial catches come from the Chesapeake and the Gulf region. Kingfish is quite oily and sometimes has pinkish flesh, although normally the flesh is white.

There are substantial resources of kingfish, which when caught is generally a by-catch. The fish ranges between 12 ounces and three pounds. It is mostly sold fresh, sometimes whole, but more commonly dressed or pan ready. It is a palatable fish, inexpensive and widely accepted.

KINGKLIP *Genypterus spp.* These fish are abundant in all southern oceans. Resources are fished from Chile, New Zealand and South Africa. The flesh is firm and white and can be used as an alternative to orange roughy. Imported fillets are sometimes available fresh.

KIPPERED FISH In the United States, this term describes fish that is brined and hot smoked. Sablefish, salmon and shad may be treated this way. The flesh is firmer and less delicate than a cold smoked fish, but has excellent smoky flavor. See SMOKED FISH.

KOSHER Food approved under the dietary laws of the Jewish religion. To meet the requirements, fish must have fins and scales. This excludes all shellfish, as well as fish such as shark, which do not have regular scales. Food has to be prepared according to strict rules, usually under the supervision of a rabbi. For a list of finfish species that are not acceptable under kosher laws, see *Old Laws in a New Market* by Joe and Carrie Regenstein, published by New York Sea Grant.

There is no particular problem in packing fish to kosher requirements and the market is very substantial.

KRILL Shrimp-like animal plankton, found in vast quantities in certain parts of the world's oceans. The North Pacific has large quantities of krill. Some species reach two inches in length. The largest, *Euphausia superba*, grows to 4 inches.

Although the meat tastes rather like shrimp, krill is very small and extraction of meat in usable form has proved difficult. Paste products are possible, but there is no very substantial market for these. The huge resources in the Antarctic and elsewhere continue to stimulate interest, but krill are unlikely to figure in offers of chilled seafoods.

L

LACEY ACT This is a federal law which makes it a felony to break any state, tribal or foreign law relating to taking, possessing, trading or selling any fish. For example, importing a fish that is legal in the United States but was fished with illegal gear in the country of origin is an offense under this law, which is comprehensive and tough.

LADYFISH *Elops saurus.* This is an important gamefish. Texas prohibits commercial fishing. The name tenpounder appears to refer to its fighting qualities rather than the size, since fish over five pounds are uncommon. Although most fish come from Florida and it is common through the Caribbean, it is also found in the Pacific.

Ladyfish is long and thin: like a herring in shape with a large, forked tail. Occasionally eaten, it has soft flesh and is very bony. It is made into a paste in certain ethnic cuisines in California, where it is sold frozen more often than fresh. Although usually shipped whole or dressed, headless fish are available and may be labeled skipjack.

LAKE PERCH *Perca flavescens.* A small northern freshwater fish with reddish skin. It is sold under many unapproved names, including yellow perch and striped perch. Whole fish are sold pan ready and are usually fried. Fillets are popular throughout the midwest, but most fish are too small to fillet, so supply is limited. Ocean perch and Pacific rockfish are common substitutes.

LING *Molva molva.* A large relative of the cod from the North Atlantic. It has firm, sweet flesh but is not very common. Kingklip is called ling in New Zealand. Cobia is sometimes called ling, incorrectly.

LINGCOD *Ophiodon elongatus.* Lingcod is not a cod. Neither is it a ling. In fact, it is a greenling.

The lingcod is caught along the entire Pacific Coast and there is a large resource in Alaska. Fish of 40 to 60 pounds are frequently landed. It is an important sport fish in Washington, Oregon and California.

Although the raw flesh sometimes has a greenish tinge to it, when cooked it is entirely white and has an excellent, sweet flavor. It is sold dressed, headless and in fillets. It is also sold steaked for barbecues and as a substitute for halibut. Lingcod has spots on the skin which distinguish it from halibut clearly enough. Although supplies are erratic, it is a good fish in use and makes excellent fish and chips.

LOBSTER *Homarus americanus.* For spiny or rock lobster, see LOBSTER, SPINY.

This species, found from Maryland northward to Labrador, and the European lobster *Homarus gammarus,* which is found in the eastern North Atlantic from Norway south to Portugal, are the only true lobsters and they are virtually identical species. Only these animals should be called lobster and calling anything else lobster is illegal in most of New England.

Lobsters grow as large as 45 pounds, though most now harvested are under four pounds. Shell color varies from brown to blue. There is no significance in the shell color. The true lobster differs from the spiny lobster in having two large claws and a much smoother shell. Fresh and cooked when alive, both are delicious. True lobster does not freeze well at all, whereas the spiny lobster freezes better and is ubiquitous on menus as lobster tail. Frozen lobster tails are consumed in far greater quantities than true lobsters. When frozen, true lobster loses a great deal of flavor and is generally rather tough.

Lobster meat is available in small quantities as a fresh product, usually from dealers who cook their weak lobsters rather than let them die and lose them totally. Fresh cooked lobster meat must be kept in plastic containers and be very carefully iced so that it stays as cold as possible without freezing. For dishes such as lobster meat salads or newburgs, it is generally much cheaper to buy lobster meat than to cook and pick lobsters yourself. Most commercially available lobster meat is frozen in cans imported from Canada, where the conservation laws permit the taking of lobsters which are too small for the live lobster trade. These are cooked, picked and frozen in cans of various sizes and formulations, generically known as cold pack lobster meat because the cans are not heated and sterilized but simply sealed and frozen. This frozen product is fine for cooked dishes. It is less desirable for salads as it tends to be tougher and not as flavored as fresh lobster meat.

Fresh lobster has to be kept alive until needed for cooking. Dead lobsters should not be used and it is an offense under many state health codes to possess dead lobsters. The problem is that when the animal dies the stomach continues its digestive functions, but digests the lobster instead of stomach contents. Byproducts of the digestive process may be both offensive and dangerous to humans. Some dealers remove and sell the tails of dead lobsters, sometimes cooked and sometimes raw. They argue that the tails are unaffected by the digestive enzymes for many days, so they are quite safe. This is possibly true, but the practice is illegal in some states, under conservation rules that prohibit the possession of lobster parts if not under health rules. Check both health and conservation codes carefully before offering lobster parts for sale.

Live lobsters are graded commercially as follows:

Chix or eighths	up to 18 ounces
Quarters	over 18 to 20 ounces
X-halves or Phillies	over 20 to 22 ounces
Halves	over 22 to 24 ounces
Three-quarters	over 24 to 28 ounces
Deuces	over 28 to 32 ounces
2½	over 32 to 40 ounces
2 to 3lb	over 32 to 48 ounces
3 to 4lb	over 48 to 64 ounces
jumbos	over 64 ounces

Some dealers have larger definitions for jumbos. Sizes quarters through deuces are called selects and are the most demanded size range. Lobsters lacking one claw are called culls and are usually graded only as chicken, select and large. Bullets lack both claws and are similarly graded.

Lobsters are often shipped in crates, ungraded or roughly graded into chix, selects and jumbos. Port dealers may ship this way to wholesalers who will undertake the work of careful and exact grading, which requires experience as well as the weighing of many of the individual lobsters. Roughly graded crate stock is generally a good buy if you have no particular size constraints for your market. However, this advantage seldom applies to anyone except larger wholesalers, who have buyers for all sizes. Restaurants and markets, as well as the distributors supplying them, generally can use only certain ranges of sizes. The high price of lobsters makes jumbos prohibitive as a retail item, so ungraded crates are of little use to retailers.

It is believed that a lobster takes about seven years to grow to one pound. It takes a further four years to grow each additional pound. Therefore, a four pounder is nearly 19 years old, which is rather ancient for such a vulnerable sea creature. The really large lobsters, 20 pounds and over, have been virtually fished out. Very few are seen any more and it will necessarily be many decades before there is a chance of many of these monsters being found again—and that is dependent on successful conservation measures leaving enough small lobsters free to grow to these great sizes and ages.

Once upon a time, Newfoundlanders were reportedly sick of eating the plentiful lobsters they could pick up on the beaches, so they fed them instead to their pigs. A.J. McClane in his *New Standard Fishing Encyclopedia* reports that in 1880 lobsters cost one penny each and about 50 million lobsters (probably over 100 million pounds) were produced. In the 1960s and 1970s landings averaged less than 30 million pounds, but have now recovered to over 50 million pounds. Maine and Massachusetts supply three quarters of the domestic catch. In most years, more live lobsters are imported from Canada than are produced in the United States.

Minimum sizes of lobsters have been increasing on a schedule begun in 1983. At the beginning of 1991, the minimum federal size increased to 3^{9}/$_{32}$ inches. At the beginning of 1992, the minimum size will reach 3^{5}/$_{16}$ inches, which is the final increase. States have the option of mandating different sizes for lobsters caught within their three miles of territorial seas. In early 1991, the northeastern states, with the exception of Rhode Island, delayed the size increase for one year. To be certain, check with the state fisheries authorities. Lobsters are measured from the back of the eye socket to the end of the carapace. The relationship between length and weight varies: lobsters moult as they grow; a newly moulted animal will be light in relation to its length. Lobster traps are designed to allow small lobsters to escape. Possession of female lobsters carrying eggs on their bodies (which is the stage prior to hatching the eggs) is strictly prohibited.

In Canada, conservation depends on limiting the seasons when each area may be fished. During the season, any size of lobster may be taken. The small ones that are under size for restaurant or home consumption are canned and frozen as lobster meat. The Canadian system seems to work as well as the American one. Be aware that in the United States possession of undersized lobsters is illegal, whether they were legally caught in Canada or illegally caught in the United States.

SOFTSHELL LOBSTERS: Lobsters grow inside their shells. The shells do not grow. When the shell becomes too small, the lobster bursts out of it, a process called moulting. After it moults, the lobster has a soft, new shell which over a period hardens into a proper shell large enough to take the animal's next stage of growth. While the new shell is hardening, the lobster is particularly vulnerable to predators. Few lobsters are caught at this stage, since they tend to hide away until they are better protected. However, many lobsters are caught, mainly in the warmest summer months, when the shell is almost hardened up. These lobsters look large for their weight: because the new shell has room for them to grow, it is not full of meat. Soft shell lobsters you are likely to receive in normal trade are not really soft, but because the shell is not full they can be squeezed and they feel softer than regular lobsters. The quality of the meat tends to be a little watery, perhaps because of the effort the lobster puts into splitting its old shell and growing a new one. Not only are they less appealing to eat, but softer lobsters are weaker and mortalities will be higher no matter what method you use to store them. For these reasons, a newly moulted or soft shell lobster is less desirable than a hard shelled animal that moulted longer ago. You are bound to receive some, but you should do what you can—which is not much—to get as few as possible.

Lobsters that have barnacles growing on the shell are nearly always full of meat, because the barnacles take a while to settle and grow. Many people reject lobsters with barnacles, because they look unsightly, but these are often the best choice. Very large lobsters, which spend many years between moults, may be covered in barnacles and resemble walking coral. Such monsters are invariably packed full of excellent meat.

POUNDED LOBSTERS: Lobsters are mainly caught in traps—lobster pots—which are baited to attract the animals, which then cannot escape from them. Usually, the traps are constructed with slats so that undersized lobsters can escape through the gaps between the slats. Most lobsters are caught close inshore and most fishermen have very small boats which can work only in comparatively calm weather. Consequently, nearly all the lobsters are landed in the months of better weather. In winter the weather prevents much fishing from small boats in the northeast, when many fishermen haul their boats and lay them up until spring.

In order to supply some of the market demand in the winter, lobsters may be kept in pounds, which are large ponds constructed along the shore. Lobsters are put into these when they are plentiful (and less expensive) during the summer so that they can be taken out and sold in the winter months. In theory, it is possible to remove the lobsters from the pounds at any time. In practice, the pounds quite often freeze over and it is impossible to drag for the animals. Other natural disasters, especially storms which damage the pounds and allow the lobsters to escape, or diseases which can become epidemics because of the concentration of lobsters in one small place, make pounding a most risky enterprise. It is not unusual for pounds to retrieve fewer than half the lobsters that are put in. The pound owner, on top of all these problems, must also gauge when to sell all his lobsters: if he leaves it too late in the year, there are freshly caught ones available and the market prefers these because they have not undergone the stress of being in the pounds. On the other hand, if the pound is emptied too quickly, he risks upsetting customers who have relied on him for off season supplies and also of missing the highest prices, which come generally just before the new season's supplies become available. Pounded lobsters, because the pounding process is certainly stressful, probably are less hardy than regularly caught animals. However, at the time when they are most available, there are few lobsters available direct from traps in the sea.

Most pounds are in Canada and there are some along the Maine coast. Pounding of lobsters is one of the more nerve-wracking gambles in the fish business.

AQUACULTURE: Many attempts have been made to grow lobsters in artificial conditions. Problems such as cannibalism, slow growth rates and poor conversion rates of feed into lobster are gradually being solved. Some apparently successful efforts are reported, mainly from the Pacific Coast, where there are no diseases which might affect lobsters, since there are no lobsters there naturally. The general pace of technological advance in aquaculture leads to the expectation that lobster culture will be economically as well as technically viable some time in the distant future.

BUYING: To summarize: when you buy lobsters, you are looking for hard shells and for animals which appear to be in good condition. Lobsters that have been kept in warmer conditions will appear livelier than ones that were kept cold, but a lively warm one may be a better bet than a sluggish cold one; it all depends and experience is the only way to be reasonably sure.

The key to profits from lobsters is being able to sell them all before they die—in other words, low mortalities. Good storage and tank systems together with careful handling are vital. It is also important to use suppliers who are similarly equipped. Lobsters caught inshore and landed within hours will be less stressed than those landed by bigger boats on longer trips. Lobsters from Rhode Island and Massachusetts are likely to have been out of the sea less time than lobsters from Canada, especially from Newfoundland. Yet the most important factor generally is the quality of the handling that both you and your suppliers provide.

CUTTING COSTS: There is little you can do to cut your raw material costs with lobsters. Larger lobsters have a higher proportion of meat, so serving half of a deuce is a better deal than a whole quarter. However, it probably has less customer appeal. Culls are also cheaper, but have limited appeal. Buying direct from the fishermen, to save the distributor's margin, often finishes up more expensive because of the organizational and administrative costs, not to mention the difficulty of maintaining supplies throughout the seasons. You will do better to concentrate on minimizing mortalities.

Lobsters are normally size graded (in the form described above). If you have a tank holding system, each size should be in a separate tank, or in a partitioned piece of a tank. There is no point in buying graded lobsters and mixing them. You either have to regrade them or lose the benefits of the grading, which, since prices vary substantially for different sizes, are very important.

INVENTORY CONTROL: An important source of losses in lobsters is theft. The simple way to monitor and reduce this is to keep inventory records in numbers, not in weights. Because most seafood inventories are kept by weight, many people still keep their lobster inventories the same way. Lobsters eat their dead companions in the tanks and also lose weight over time, since they are not feeding (except on their dead companions). It is impossible to match up incoming and outgoing weights. Keeping lobster inventory records by weight makes it easier for people to steal lobsters without being detected.

Because you should be keeping each size separately, it is simple to keep inventory records in terms of numbers of lobsters. Apart from any that die and are completely eaten by the others—which is rare, firstly because you should have been picking over the tanks for dead lobsters frequently so there is not time for a corpse to be fully consumed, and secondly because there usually is enough left of the claws and head to identify that there had been a lobster. The numbers of lobsters that go out should always equal the numbers that came in.

Because your control is certain, employees are deterred from larceny. Dishonest practices, such as putting extra lobsters in a box for the truck driver to sell for cash during his deliveries or going home with a couple of lobsters concealed, are less attractive if employees know that you will be looking for the missing pieces and that they are prime suspects.

It is usually quite feasible to count the stock whenever you drain the tanks for cleaning and this is an additional security. This particular count should be made by administrative personnel, not by tank room staff.

COOKING: Mention has been made already of cooking lobsters that look as though they are about to die. This should be a daily or twice daily routine, depending on volume. Boil salted water, add lobsters for ten minutes for the first pound plus one minute for each additional ¼ lb. Two pound lobsters should therefore be cooked for 14 minutes. Once cooked, the lobster should be plunged into iced water for as long as the original cooking time, unless it is to be eaten hot, immediately. This ensures that the meat will not continue to cook inside the shell. Make sure that the iced water is changed for each batch of lobsters. Cooking live lobsters for use or sale as whole cooked lobsters requires the same techniques as cooking the weak ones. Cooking should be timed from the moment the water returns to the boil. If steam is used to heat the water, it is advisable to use a thermometer to check when the water actually boils, since the steam makes the water bubble and appear to be boiling long before the temperature reaches 212°F. Be cautious about cooking too many lobsters at one time, as this cools the cooking water too much. The lobsters then cook rather slowly and a tough product results.

Large lobsters take longer to cook through than small lobsters, so it is important to ensure that batches of the fish that are being cooked together are fairly uniform in size.

Irrespective of the size of the lobsters, they will lose weight during cooking. The percentage loss varies considerably. It may be as little as 5 percent or as much as 25 percent. The average is probably around 15 percent. Weight loss depends on many factors, such as the condition of the lobster and the cooking technique. It is difficult to forecast what will happen, except that some weight will certainly be lost.

Cooked lobsters may be sold as whole, cooked lobsters or picked and sold as fresh lobster meat. Attractively garnished displays make both more saleable in markets and in restaurants. Although these are salvage products, they can still be profitable and earn money above costs. Shelf life of cooked lobster meat is limited. It is not advisable to sell it after more than three days. Cooked product is much more vulnerable to bacteria growth than raw product (for the same reasons that cooked product is easier to eat than raw product) and must be handled with scrupulous attention to hygiene. Listeria is a particular hazard with cooked lobster meat. It is even found on imported frozen canned product. Cooked and raw product should never be displayed together in a showcase, especially where drip or bacteria from raw food might be transferred to the cooked items. Despite the attractive appearance of mixed displays, this is something that should never be done unless you are sure that the display is a promotional expense and the product will not be sold. This applies to all seafoods, not just to lobsters.

An occasional problem may occur when consumers complain of the greenish meat in some lobsters. If properly cooked, this part, the liver or tomalley, should turn red and is good to eat, but the under-cooked version is not appetizing. Cooking depends on the temperature reached, not on the cooking time. At boiling point it turns red in a couple of minutes. At 160°F it may take 30 minutes to turn red, and at lower temperatures, even though the surrounding meat is cooked, it may not change color at all. It is therefore important to make sure that the cooking water regains a fast boil in a short time.

If you must freeze lobsters, the product is better if the lobsters are cooked before freezing. Frozen, raw lobsters usually have little taste and the texture of the meat, which is difficult to remove from the shell, is also frequently unappealing.

LOBSTER HANDLING AND STORAGE: Gentle care is the key. Lobsters are delicate and they are out of their natural element. It is very easy to weaken or kill them by throwing them around, by dropping or

LOBSTER

banging their containers or by keeping them too warm or too cold. Lobster mortalities are very expensive, and have to be avoided if buying and selling of lobsters is to be worthwhile.

Tank systems with flowing water are preferred for storing lobsters. These may use pumped sea water, if the installation is close enough to the shore, or recirculated natural or artificial sea water. In many respects, artificial sea water is better: it is consistent in chemical composition, while natural sea water varies according to the tides and the weather. Also, natural sea water may import diseases and frequently imports small mussels, barnacles and similar creatures which thrive in the tanks and pipes and have to be removed with considerable effort.

There are a number of manufacturers of tank systems. They can advise on suitable types of tank, filtration and salts. Modern systems are pretty reliable, if the user follows the instructions. *Live Holding Systems* by Peter Lappin (see Bibliography for details) gives full details for operating lobster tanks.

The lobster requires plenty of oxygen in the water. This is usually provided by pumping or cascading the water so that it mixes with air. However, it is possible to add too much oxygen to the water, so it is important to use the right pumping rate.

Lobsters generally live longer if they are kept cold. The ideal temperature for the sea water in a tank system is probably around freezing point, say between 30°F and 35°F. At this temperature the lobsters are virtually dormant. They move very little and use hardly any of their stored energy, which means they maintain their weight. The major drawback is that they become rather brittle, with a tendency to shed legs and even claws when they are moved from such cold water. Careful handling can solve this problem, but most tank suppliers recommend rather higher temperatures. For longer term storage, if you have such a need, lower temperatures are better.

If lobsters arrive too warm, they should be chilled in the cooler, not put straight into the tanks where the shock of sudden temperature change could kill some of them. Under all other circumstances, lobsters should be put into the water as soon as possible.

SHIPPING: Transportation of lobsters should also be done with care. Although 50 pound boxes are often used, 25 pound boxes are preferable because there is less crushing and therefore less stressing of the animals. Small amounts of ice may be added to the boxes, to ensure that the gills stay moist, but too much ice may suffocate the lobsters: their gills are designed for salt water and do not work in fresh water. If you have access to wet seaweed, this is an excellent insulation and wetting medium to use when shipping lobsters. Some airfreight dealers still like to use this whenever possible.

Weights and counts should be checked as soon as you receive lobsters.

No supplier should entertain a claim made after his truck has left the scene. Once you have signed for the shipment, the dead ones are yours. It is also wise to agree on a mortality allowance with the supplier beforehand: some deaths are inevitable. Guarantees of 100 percent live lobsters are unrealistic.

Lobsters should never be thrown around, dropped, shaken or otherwise abused. Remember that they are alive and under stress and that stress kills them. Lobsters represent a lot of money and should be treated with care and respect.

LOBSTER, SPINY

Panulirus spp., Palinurus spp. and *Jasus spp.* These are also called rock lobster, Florida lobster, lobster tails, crawfish and crayfish. The species most commonly landed in Florida is *Panulirus argus.*

Spiny lobster or rock lobster (the two terms are interchangeable—there is no distinction between them) is a crustacean similar to the northern lobster but lacking the claws. In the United States, most fresh commercial supplies come from Florida, but Hawaii and California also have resources. In most years, over 90 percent of total spiny lobster supplies are imported, mainly as frozen tails from Australia, New Zealand, Brazil and many other countries. Frozen lobster tails are a major seafood commodity and most Americans outside the Northeast tend to think of these tails first when lobster is mentioned.

Live spiny lobster should be handled in much the same way as live northern lobsters. They are at least as delicate. Most of the comments about examining, keeping and using lobsters apply also to spiny lobsters.

Florida, which officially uses the name crawfish, prohibits catching lobsters from April 1 through July 25, to protect breeding fish. Minimum size is also limited, by both Florida and the federal government. Florida requires that the fish have to be landed whole, which makes it possible to police minimum size regulations: if only tails are landed, it is difficult to prove how big the lobster was. Nearly three quarters of the harvest is landed between August and November.

About half the Florida catch is cooked and frozen. This means that taking into account the small catches elsewhere, only about 7 percent of the market for spiny lobsters is supplied by fresh product. Much of that is consumed in Florida.

For information on the many species of spiny lobster and how to identify them, see *Lobsters of the World, An Illustrated Guide* by Austin Williams, published by Van Nostrand Reinhold/Osprey Books.

LOBSTERETTE

Nephrops norvegicus. Also called Norway lobster, Dublin Bay prawn, langoustine, Icelandic baby lobster, deep sea lobster, scampi, dainty tails, deep sea dainties and probably other names too. *Nephrops andamanicus* is a similar animal from southern waters and *Metanephrops rubellas* is another similar animal sometimes imported from Brazil. There are a number of other species worldwide. Langostinos (*Cervimunida johni*), which are imported from Central and South America, are entirely different animals. Lobsterette is not much used in trade, but it is favored by AFS scientists and since it is clear, descriptive and not confusing, it may well be adopted by the FDA.

The original lobsterette, or langoustine, or Dublin Bay prawn, or scampi, is a small crustacean like a small lobster found mainly in northern European waters and the Mediterranean. It is a close relative of the clawed lobster. Fresh, this is one of the most delicious seafoods. Frozen, it is also pretty good. Langoustines are generally around three to four ounces, giving edible tails of about one ounce. They will grow as large as two pounds. Large langoustines are seldom seen.

Because of its popularity, scarcity and high price, the name is applied to any sea creature with an approximate appearance from all over the world. Different names manage to obscure rather than clarify just what these items really are.

Fresh langoustines are occasionally flown to the United States from Denmark and Iceland at enormous prices. They look like very large shrimp, with two long claws which are about the same size. The color is orange pink and the flesh is very white when cooked.

LUMPFISH

Cyclopterus lumpus. Also called, incorrectly, lumpsucker, snailfish, henfish and other names. The only approved name in the United States is lumpfish. It is a large (20 pound) but slow growing North Atlantic species, not normally used for food in the United States. The roe is used for the production of a dyed and inferior imitation caviar.

M

MACKEREL. *Scombridae.* Scientifically, this family includes a total of 48 species of mackerels and tunas. Mackerels, Spanish mackerels and tunas are closely related. Commercially, the distinctions are defined differently. Jack mackerel is a member of the jack family, but is regarded as a mackerel. Tunas and mackerels are not in any way interchangeable for marketing purposes. This section follows accepted commercial definitions of mackerels. ATKA MACKEREL is covered under its own heading (see above).

ATLANTIC MACKEREL: *Scomber scombrus*. Small mackerel are sometimes called tinker mackerel.

A small, oily fish, found on both sides of the North Atlantic. Mackerel are usually between 1 and 2 pounds and up to about 18 inches in length. Larger fish, up to four pounds and 22 inches, are seen. The skin is thin and smooth, with tiny scales. The brilliant greenish blue and silver iridescent colors are characteristic, although they fade after the fish dies. The meat is brownish, but when cooked is creamy. It has a strong flavor.

Mackerel was once a staple food in Europe and salted mackerel was an important item in 19th century New England, being replaced in importance by canned mackerel during the last quarter of the century. Mackerel resources now appear to be far smaller than they once were and the fish is less highly regarded. It is, however, a very cheap and excellent fish if properly handled and deserves more consideration for its potential value as a fresh seafood.

Mackerel are a shoaling fish and, when they are available to the fishermen, are caught in huge quantities. On the U.S. and Canadian side of the Atlantic, they are at their best in fall, when the fat content is highest. The fish feed little during the winter and spawn in spring, after which they are thin and of poor quality. They fatten up rapidly throughout the summer and fall. For almost every purpose—canning, smoking, salting, pickling or eating fresh— mackerel have the most flavor and the best texture when they are fattest.

Partly because mackerel are caught in such large volumes, they are not gutted on the boats and handling may be less careful than it should be. It is vital to put mackerel in ice as soon as it is landed. It loses flavor very quickly otherwise. If properly iced and kept cold, mackerel has a surprisingly long shelf life. Because it is an oily fish, mackerel will turn rancid quickly if not properly handled. The fish are invariably offered whole. The skins should retain some brightness and be tight. Loose, wrinkled skin is a sign of aging. Look for bright, concave eyes and reddish gills. Fillets with the skin on may also be available. The flesh should not be gaping.

Mackerel must be handled with care. Ice and cold temperatures are essential and, since the fish tend to drip oil into the ice, changing the ice frequently is also important. Ungutted fish has a tendency to "blow"— gas extends the stomach as the contents of the viscera decompose—and if it is packed too high, the weight of the fish on top will cause the bellies of some of the fish underneath to burst, which is unattractive and wasteful. The bacteria in the viscera will also decompose the mackerel flesh it contacts, rapidly reducing the value of the shipment.

Fresh mackerel are excellent broiled. They are even better hot smoked, resembling smoked trout but far less costly. Mackerel is highly versatile. If good, fresh supplies can be found and maintained, the fish will repay the effort involved in handling it. Resources appear ample; the market is lacking.

CERO: *Scomberomorus regalis*. Called pintada in parts of Central America (and the English translation painted mackerel in parts of the United States). Similar to the Spanish mackerel, found in quantity in the warmer parts of the Atlantic, especially around Florida, the Caribbean and northern South America, the cero commercially averages between five and 10 pounds, and may reach over 50 pounds. Comments under Spanish mackerel (see below) apply to cero.

FRIGATE MACKEREL: *Auxis thazard*. See BONITO, which is a different but very similar fish. Commercially, the two are handled and treated in the same way.

HORSE MACKEREL: *Trachurus trachurus*. An Atlantic species similar to the Pacific mackerel (see below). The term is also used, incorrectly, for king or other large mackerels on the Atlantic coast. Bluefin tuna, which are totally different, are sometimes called horse mackerel.

JACK MACKEREL: *Trachurus symmetricus*. A Pacific species, generally around one pound, important in California where much of it is canned. It is also good smoked. It is prized for swordfish bait. Although technically a jack, not a mackerel at all, it is very similar in appearance and use to Pacific mackerel. The fish is abundant in many parts of the Indo-Pacific basin.

There are several very similar *Trachurus* species which are not distinguished commercially. Some of these may be called scad.

KING MACKEREL: *Scomberomorus cavalla*. This fish is sometimes called cavalla or kingfish. The FDA frowns on these names.

Found throughout the warmer Atlantic, and plentiful around Florida between November and March, king mackerel are large, up to 100 pounds, although 10 to 20 pounds is the usual size. The meat is well flavored, but it has more fat than the Spanish mackerel and is best used for broiling or smoking. Commercial net fishing for the species has been phased out. Line caught fish can be bled on capture, which improves the quality.

Because of its popularity as a game fish, the species may largely disappear from commercial use as more of the resource is reserved for recreational fishermen. However, fresh and frozen supplies are available from Mexico, Brazil and other countries.

King mackerel is offered headless and dressed, or it may be steaked. Only very large fish are filleted.

PACIFIC MACKEREL: *Scomber japonicus*. Very similar to the Atlantic mackerel, the Pacific mackerel is an important resource from the Gulf of Alaska to Mexico. It is found throughout the Pacific. The species is found in much of the Atlantic as well, where it may be distinguished by the name of chub mackerel. Since it is a little smaller than Atlantic mackerel, it may also be sold as tinker on the market.

Pacific mackerel is generally lower in oil content than Atlantic mackerel, which makes it taste somewhat drier. Most of the comments made about Atlantic mackerel apply also to this species.

SIERRA: *Scomberomorus sierra*. A Pacific variant of the Spanish mackerel (see below), found mainly along the Mexican coast.

SPANISH MACKEREL: *Scomberomorus maculatus*. Regarded by many as a top quality table fish, Spanish mackerel are generally between two and three pounds and are plentiful around Florida between November and March. Later in the year, quantities are caught by sport fishermen as far north as Rhode Island. However, Florida in most years accounts for 90 percent of the total commercial catch, three quarters of it from the Gulf coast. The resource appears to be in good condition, but pressure from sport fishing interests may reserve the domestic fishery for recreational fishermen within a few years. There are several very similar

MAHIMAHI

Spanish mackerels fished around the world and imported into the United States.

Spanish mackerel is easily distinguished from king mackerel and cero because it has golden spots on the sides and none of the lateral stripes which other mackerels have.

The meat is quite white and is very well flavored. It can be used for baking, broiling, smoking and in most other ways. Spanish mackerel is often cheaper than king mackerel. Small fish around one pound are usually sold whole or dressed; larger fish may be filleted.

MAHIMAHI

Coryphaena hippurus. Firstly, it is essential to make clear that this is a fish, not a mammal. It is unfortunate that it is often called a dolphin, but it is not related in any way whatsoever to the protected mammal dolphin, which is a porpoise (a small whale species). Mahimahi, the Hawaiian name, is a far better description, avoiding any possible confusion and argument about what is being offered. Regrettably, the FDA permits the name dolphin in interstate trade. The Spanish name is dorado.

Fresh mahi comes from both the Atlantic and the Pacific. Most catches are made in the summer months as the fish follow warm currents. Long popular in Hawaii, mahi availability in the form of cheap imported frozen fillets created markets in California and Florida. It is now eaten widely throughout the United States. Frozen supplies are available from many parts of the world, including Taiwan and Ecuador. Fish farmers are experimenting with mahimahi, which has given good results in trials.

The fish has unusual and magnificent coloring on the skin—bright blues, greens and yellows which make it most distinctive. Although the colors fade after the fish is caught, enough remains for the skin to be an attractive part of a buffet display. Consequently, although the skin is tough and not really edible, mahi is invariably sold and cooked with the skin on.

Size varies greatly, from a couple of pounds to over 50 pounds, so the size and thickness of fillets is also very varied. The reddish meat turns white when cooked and has a large flake. Flavor is better if the fish is bled by cutting off the tail as soon as it is caught. Frozen imported mahi from Ecuador, Taiwan and other countries is widely available. Fresh mahi is substantially better. Watch for yellowtail instead of mahimahi when the price of mahi is particularly high. Fresh product is available filleted (sometimes called loins) or headless dressed, sometimes with tails and collar bones removed.

Mahi flesh is associated occasionally with histamine poisoning. See HISTAMINES for details. Since this risk is aggravated by poor handling, it is important that mahi are cooled and iced as soon as possible after capture and are properly treated through the distribution chain.

MELANOSIS   Also called black spot. Melanosis is the blackening of the shell of crustaceans during storage and shipment. Shrimp and some species of crab are particularly prone to developing black spot. It is caused by natural pigmentation in the skin membrane just under the shell. This pigmentation seeps out and causes black spots and patches on the shell. Melanosis is in no way harmful, simply unpleasing. Severe cases may mark the meat as well as the shell. Again, this is not harmful.

Melanosis is much more likely if product has been poorly handled, especially if it was left in warm temperatures too long after being caught. It is not very likely on fresh product, since this has to be consumed fairly quickly, but can be a problem with frozen product, especially whole shrimp. Freezing does not prevent melanosis.

Use of certain chemicals can delay black spot. Most of these chemicals are illegal in the United States.

MENHADEN   *Brevoortia tyrannus*. The fish is sometimes called pogy or mossbunker. Do not confuse pogy (menhaden) with porgy, which is scup. Menhaden are small, herring-like fish which support one of the largest and most important fisheries in the United States, along the Atlantic and Gulf coasts. Menhaden are extremely oily and not generally regarded as palatable, but enormous quantities are used for fishmeal and fish oil. It is also a popular baitfish for bluefish and tuna fishing.

It is possible that surimi could be made from menhaden.

MERCURY   A heavy metal which occurs naturally in some seafoods and tends to accumulate in the flesh. Animals which live a long time, such as swordfish, tuna and dogfish, may accumulate significant quantities.

Federal rules now limit mercury to one part per million (1 ppm), which conforms to rules in most countries which trade seafoods with the United States.

In general, smaller fish will have less mercury than larger ones, because they are younger and so have had less time to build up mercury to a high level. Swordfish is generally acceptable up to about 150 pounds and many fish which are much larger are also below the level.

The mercury problem has been exaggerated in the past. It occurs naturally, but in relatively few seafoods.

MERUS   The section of the leg of a king or snow crab which is closest to the shoulder. It is the largest and thickest leg segment. Merus sections and merus meat command premium prices.

MILKFISH *Chanos chanos.* A very large (up to 50 pound) fish shaped like a herring. The species has been farmed for centuries in ponds in many parts of Asia. Milkfish cultivation is starting in Hawaii.

The flesh is pink and rather soft, but milkfish may become more common in the United States. It can be used for sashimi, though fish cakes and fish balls may offer a better outlet for larger quantities.

MILT Male gonads, the equivalent of roe in the female. Herring milts are eaten and well regarded in many European countries, where supplies from Canada of frozen milts, also known as soft roes, are an important commodity.

MOLLUSC Also spelled mollusk. Shellfish are either molluscs or crustaceans. Shells are the equivalent of the skeleton in higher animals, so shellfish have no bones in the meat. There are three types of molluscs:

Bivalves have two shells that enclose all or most of the animal when closed. Clams, oysters, mussels and scallops are bivalves.
Univalves have a single shell. The animal lives inside the shell and often has a shell-like plate which can be closed over the opening. Snails, conch and abalone are univalves.
Cephalopods do not usually have an external shell. The name indicates feet growing out of the head. Squid, octopus and cuttlefish are the cephalopods commonly eaten. These have a piece of cartilage inside the body which serves as the skeleton. It is called the pen or quill in squid and the bone in cuttlefish.

MONKFISH *Lophius americanus* and *Lophius piscatorius.* The names monkfish and goosefish are approved by the FDA. The fish is also called anglerfish, frogfish, sea devil, ocean blowfish, bellyfish, frogfish and allmouth. Found throughout the world, but mainly supplied from the North Atlantic, the monkfish is a large, deep sea fish without scales (therefore not suitable for Kosher markets) and of outstanding ugliness. Fishermen cut off the tail portion, throwing the huge head and body section back into the sea. Monkfish as large as 75 pounds have been caught, but most fish are much smaller, usually less than 15 pounds.

Tails as landed have a number of layers of skin and membrane, which are usually removed by peeling them back from the cut (head) end. Monkfish has a single bone through the tail and two boneless fillets are easily removed from this. Tails may be sold whole or as skinless fillets, but however you buy it the skin and membranes should be removed before use.

The flesh is white and firm with a mild flavor. Because of its texture it can be used to supplement lobster meat in some recipes.

Monkfish livers are airfreighted fresh to Japan. Since these are often worth more than the meat, the state of this market may affect domestic supplies of fresh monkfish tails and fillets.

MOULTING Crustaceans grow inside their shells. When they become uncomfortably large for the shell, they burst out of it and develop a new shell. This process is called moulting. Soft shell crabs are blue crabs which have just moulted. (See under CRABS). Although soft shell crabs and soft shell crawfish are great delicacies and important products, other moulting crustaceans are less appetizing. Soft shell lobsters have soft, watery and tasteless flesh, for example.

MULLET *Mugil spp.* This is not the same fish as the red mullet used in French and other Mediterranean cooking: that fish is actually a GOATFISH and is seldom eaten in the United States *Mugil cephalus* is striped mullet, sometimes incorrectly called black or gray mullet; *Mugil brabatus* and *Mugil surmuletus* are red mullet; *Mugil liza* is sometimes called lisa; *Mugil curema* is white mullet, sometimes called silver mullet.

Canadian suckers or buffalo, which are a freshwater fish, may be offered as mullet, which is what they are sometimes called in Canada. They do not resemble mullet in any way, the term being used to increase market appeal. Suckers are *not* replacements or alternatives for mullet. They have bland, white flesh, while mullet is distinctive and comparatively oily.

Striped mullet are small fish, up to four pounds, usually under two pounds, fished heavily during the summer and fall along the Florida coast. Heaviest landings are in early fall on the Gulf coast and late fall on the Atlantic as far north as North Carolina. Silver mullet is found only on the Atlantic coast and is most abundant in spring. Mullet is a cheap fish with rather oily flesh, although the removal of the dark lateral band extends the shelf life considerably and makes the flavor much lighter.

Female fish are prized for their roe in Japan and Taiwan and during some years enormous quantities of mullet roe are exported (though most Eastern buyers prefer to buy the whole fish and process the roes themselves). Frozen mullet is also exported in huge quantities, though irregularly, to African and Caribbean countries.

Domestic consumption of fresh mullet is largely concentrated in Florida, with some usage in the northeast and the midwest. Florida processors generally ship the fish whole and it is filleted if required by the local distributor. Because of its high oil content, mullet needs to be kept well refrigerated and used quickly or it turns rancid. Smoked mullet is said to be excellent though it is not offered very much.

MUSSEL

BLUE MUSSELS: *Mytilus edulis*. The common or blue mussel is found on both the Atlantic and Pacific coasts of the United States and is a very common shellfish throughout Europe. *Mytilus galloprovincialis*, the Mediterranean mussel, is also found on the West Coast. The two species are so similar that some scientists believe that one is a subspecies of the other, though opinions differ as to which is the main species. Commercially, the two are indistinguishable. Blue mussels are circumpolar. On the Atlantic coast of North America they can be found from the Canadian arctic south to North Carolina. On the Pacific coast, they range as far south as southern California. In the eastern Atlantic, mussels are found from the White Sea to north Africa. They have been introduced successfully to Japan as well as to China, where they are being farmed. Mussels are abundant and inexpensive.

Mussels are bivalve molluscs, that is, they have two shells which form the skeleton. Blue mussels are regularly shaped with light ridges running around the shell. They vary greatly in color from light brown through greenish to dark blue and black. Shell color has no quality, origin or any other significance. The soft parts enclosed by the shells are all edible. Meat colors also vary from creamy to brown and orange. There is no quality significance in the meat color. In the United States, mussels are normally cooked, but in Europe they are sometimes eaten raw.

Mussels have light shells. Meat yields will, in general, range between 25 and 55 percent, depending on how they were grown, the time of year and their spawning cycles. A bushel of mussels might weigh as much as 60 pounds and yield 30 pounds of meat: very inexpensive protein.

Because the whole animal is eaten and they are often cooked only lightly (not enough to be sure of destroying pathogenic organisms in the meats), mussel harvesting and distribution are subject to the controls of the National Shellfish Sanitation Program (NSSP). Mussels may only be harvested from approved water and every company handling them through the distribution chain has to be licensed by the state.

Although mussels are not a major seafood product in the United States, huge quantities are eaten in Europe and Asia. Worldwide, mussel harvests normally exceed 900,000 metric tons annually, with Spain the leading producer with over 200,000 tons. China is Asia's leading mussel producer, with production rising rapidly towards 200,000 tons a year. The United States, by contrast, produced only 21,000 tons in 1987 (the latest year for which there are comparable international statistics). Over 95 percent of American output comes from Maine and Massachusetts. Canadian production in the Maritimes is increasing, with Prince Edward Island putting particular effort into expanding its production and markets.

On the west coast, production is small and limited by conditions unsuitable for growing larger mussels and also by large populations of diving ducks, which have voracious appetites for mussels.

Most mussels produced in the United States are farmed in some way, usually by relaying them on the seabed in areas known to produce fast growth. Small quantities are grown off the bottom, on poles, lines or from rafts and these methods tend to produce the fastest growing mussels. Mussels sometimes grow pearls, which can break your tooth. The pearls grow slowly, so they are much less likely to be found in mussels which have grown fast, such as those grown from rafts.

Live mussels are generally packed in onion bags or similar sacks, holding about a bushel. It is better to negotiate to buy mussels by weight than by volume, especially if you sell them by the pound. They should be thoroughly cleaned of seaweed, barnacles and other marine detritus before they are shipped. However, it is better to leave attached the black byssal threads, known as beards. Removing these shortens the shelf life of the mussel. The beards should be pulled out before the shellfish are cooked.

Live mussels, like all live shellfish, must be kept cool and moist. They must be handled gently. The shells are thin, quite brittle and easily cracked. Once the shells are cracked or broken, the mussel does not live long. In any case, it should not be used because of the risk of bacterial contamination. Mussels should be stored and transported on pallets so that they do not sit in meltwater from ice. Fresh water kills them, so although ice is essential to keep them cool and damp, make sure that they do not sit in the melt. Bags of mussels should be kept tightly closed, so that the animals are under a little pressure. This helps to prevent them from opening. When they open they dry out and, fairly quickly, die.

Never use dead mussels (test open shells by tapping them on the table: if they are alive they will close up). Discard any that are cracked or chipped.

Under the rules of the NSSP, bags and other containers of live mussels must carry a tag showing the origin and date of harvest. Refuse any bag which does not have the tag. By law, the tags must be kept by the final receiver for at least three months. It is sound practice to keep them for longer or at least make sure that the information is entered on a permanent record, such as the invoice.

There are numerous mussel products and some are beginning to find new markets for these tasty and nutritious shellfish. Fresh shucked meats are available, as are fresh mussels on the half shell and meats in many different marinades and sauces. Fresh mussel products must be labeled with the license number of the packer. Again, you must keep records of the origin of each container for at least 90 days.

Mussels are inexpensive and readily available. Production could be expanded enormously to meet any increased demand, but Americans find mussels unfamiliar. Mussel meats can be breaded and used in

MUSSEL

fishermen's platters to cut portion costs dramatically, compared with using similar quantities of oysters or clams. Meats added to basic marinara-style pasta sauces or used instead of clams in sauce recipes will again offer lower portion costs for equal nutrition and excellent flavor and texture. Mussels do not deserve to be neglected.

GREEN MUSSELS: *Perna canaliculus*. These are also called green lipped mussels. In the United States. they are also known as New Zealand mussels. Similar species grow in other countries in the southern hemisphere. Green mussels are quite large and are usually harvested at about three and a half to four inches. The shells are attractively green. The meats look similar to blue mussel meats but, because they are generally larger, tend to taste more succulent.

Green mussels are available live, shipped by air from New Zealand. They are size graded by count or weight, depending on the shipper or the buyer's requirements.

HORSE MUSSELS: *Modiolus spp*. These are larger than blue mussels and have red meats, which are coarser and tougher than those of the blue mussel. There are resources available on both coasts, but few are exploited. Breaded horse mussel meats have found some market success in Canada.

N

NEEDLEFISH *Cololabis saira* is the Pacific saury. *Scomberesox saurus* is the Atlantic saury. The FDA prefers the name saury. These are long, silvery fish with beaks. Although popular in some European countries, needlefish are not normally eaten in the United States. They can be cut into sections, like an eel, and broiled, poached or fried. Saury are important food fish in Japan, where they are canned, making a product similar to canned mackerel. Quantities of this are exported to the United States from time to time.

NET WEIGHTS Determining the actual net weight of product received can be tricky. Fish buried in ice must be rinsed clean before it can be weighed. Fillets naturally leach fluids, so there will be some drip at the bottom of containers. There is no simple approved method for checking drained weights of fresh seafoods, although such methods do exist for checking frozen glazed seafood products. Checking weights requires some experience and common sense.

Two seafoods which suffer from weight problems and disputes are scallop meats and shucked oysters. Particular aspects of the weight problem are discussed under those two headings.

NILE PERCH *Lates niloticus*. In 1990, the FDA approved the alternative name of Lake Victoria perch. This is a large freshwater bass (it grows over 300 pounds) from Africa, with rather coarse flesh. Smaller fish provide white, bland fillets similar to tilapia. The species may be suitable for aquaculture.

NOBBING Removing the head and guts from a fish with a vertical cut across the collar so that the guts are removed through the neck opening without cutting the belly. This is one of the standard ways to process herring.

NOMENCLATURE

NOMENCLATURE Although hundreds of species of fish and shellfish are used regularly in the United States (there are thousands of edible species available worldwide), consumers recognize the names of only a few. Lobster, flounder, salmon, sole and cod are familiar to almost everyone. Saury, muskellunge, goosefish or conger eel are not widely known. Many little known fish are very good to eat, but they are difficult to sell because the names are unfamiliar.

Marketers are tempted to apply well known names such as snapper and sole to little known fish that may or may not share their favorable characteristics. This practice is, of course, fraudulent. Turbot has been frequently sold as flounder or sole. Arrowtooth has been labeled as turbot. Tilapia with pink skin has been sold as snapper. Ocean perch becomes lake perch or red snapper. Calico scallops are promoted in the marketplace as much more costly bay scallops.

The system for defining fish names in the past was unclear and rather rigid. To clarify and codify the rules, as well as to make it possible to find sensible names for the many new products that are appearing on the U.S. market, in 1988 the FDA published a small book called *The Fish List: FDA Guide to Acceptable Market Names for Food Fish Sold in Interstate Commerce*. This lists over 1,000 species of edible finfish which might be found on the U.S. market. The list gives Market Names and Common Names which are acceptable for each species. It also lists other names by which the fish is sometimes known. These other names are generally not acceptable on interstate shipments. Throughout this book, the names approved by the FDA are indicated, together with the more likely of those that are not acceptable. *The Fish List* is reproduced in the Appendix, edited for typographical errors and with an added indication of the scientific family and area of origin for each species.

The document also addresses the problem of how to establish a name for a species that is not listed. The procedure is to examine the fish in the light of a set of priorities and to choose a name that relates to the highest relevant priority. These priorities are straightforward and are also reproduced in the Appendix. If there is no established name, you can create one (you need the FDA's approval). It is not possible to trade on the name of another species, but then if it were, every new fish would be called golden scrod snapper.

California has some locally used names, which are acceptable in the state but not in interstate trade. Helpfully, the University of California Cooperative Extension has published its own version of the *Fish List*, annotated with California's idiosyncracies (which include calling some rockfishes by the desirable name of red snapper). The publication is called *Menu and Advertising Guidelines for California Restaurants, Retailers and Their Seafood Suppliers*.

Shellfish names are less confusing and contentious. The FDA is planning to publish a Shellfish List sometime. In the meantime, the American Fisheries Society has two publications that can be used as guidance: *Common and Scientific Names of Aquatic Invertebrates from the United States and Canada—Molluscs* and *Common and Scientific Names of Aquatic Invertebrates from the United States and Canada—Decapod Crustaceans.*

Using the wrong name constitutes misbranding. The FDA is able to seize product and require it to be relabeled. Deliberate misbranding can also incur criminal penalties.

O

OCEAN POUT *Macrozoarces americanus* and *Zoarces viviparus*. The FDA permits eelpout or ocean pout for these fish. Other names you might find include yellow pout, sea pout, muttonfish and viviparous blenny. The name is unimportant because the fish is almost inedible.

Fillets from these deservedly underexploited Atlantic species are boneless, long and thin and have a quite attractive creamy color. They look nice on a retail display, prompting more adventurous consumers to try them. Ocean pout, unfortunately, is tough, chewy and tasteless once it has been cooked. It has been reported that if the fillets are passed through a meat needle tenderizer or are pounded with a mallet, they are apparently greatly improved. Ocean pout can be used effectively in soups and chowders.

OCTOPUS *Octopodidae*. Cephalopod molluscs which feed on fish and shellfish. Although octopus is quite plentiful in waters along all U.S. coasts, little is eaten and most of that is supplied by frozen imports. Octopus in the United States is limited to ethnic and regional markets. North Pacific species may reach 16 feet, but most examples available to U.S. fishermen weigh less than five pounds. They are highly regarded as bait for halibut and other large fish.

Octopus is normally cleaned before distribution, which means viscera, eyes and beaks are removed. The ink sac is used in many recipes and may be left with the animal. The skin is edible.

Octopus is best tenderized by dipping it two or three times in boiling water for five seconds each time, allowing it to cool between each dipping.

Octopus is an extremely valuable seafood in many parts of the world, including Japan, Mexico and southern Europe. It is eaten raw, cooked, dried or smoked.

OMEGA-3 Also written as n-3, the term describes certain polyunsaturated fatty acids found in fish oils, which appear to have beneficial effects on human health. The health and nutrition advantages of fish and shellfish are a major marketing tool for the industry.

OPAH *Lampris guttatus*. Frequently called moonfish, though this name is not approved and causes confusion with lookdowns, which are quite different. It is occasionally called sunfish, which is quite wrong. Opah is a large pelagic fish found in most temperate and subtropical seas. It is a large fish, growing over five feet in length. It has a deep body, like a pomfret.

The flesh is red and marbled and can readily be cut into boneless steaks and portions, suitable for broiling.

ORANGE ROUGHY *Hoplostethus atlanticus*. A deep water fish from the southern hemisphere which provides a firm, white, boneless fillet with bland flavor. It is popular for retail sale in the United States. A related Atlantic species, which is not available in commercial quantities, was called slimehead by scientists. In 1982, when orange roughy was first offered to American buyers from New Zealand packers, the more attractive name was secured for this species, which had not previously been marketed.

Because the fish is caught in deep water far from packing plants on land, it is mostly frozen at sea. Some of it is thawed and filleted later, some is filleted and frozen on board the fishing vessel. Fresh orange roughy is hard to find because of the logistics of getting it from the fishing grounds to the customer in a short time. Fillets range up to about 10 ounces and must be deep skinned to remove the underlying layer of fat.

OYSTERS *Ostreidae*. Bivalve molluscs. Oysters were farmed 2000 years ago by the Romans and almost all commercial production today is farmed to some extent. Oysters generally prefer estuarine conditions where the salinity is quite low. Since estuaries have been heavily polluted, many once prime oyster grounds have been lost. As rivers are cleaned up, some of these areas again produce oysters, but the industry is constantly vigilant for pollution problems and has a vital stake in ensuring that marine environments are kept clean.

World production of oysters is over 1 million tons annually, of which perhaps 70 percent are Pacific oysters. As recently as 1985, the United States was the world's largest producer, but diseases affecting the eastern oyster and increased Asian production has pushed the United States into third place behind South Korea and Japan. Although American production is still substantial, it is insufficient to meet demand; oysters and oyster products are imported from a number of countries, including Chile, New Zealand and South Korea. By 1989, imports, chiefly of canned and frozen meats, supplied over half of the U.S. market (meat weight basis). The Pacific oyster is the most important in world terms, but in the United States the eastern oyster still accounts for close to 70 percent of production and determines the standards by which other oysters are judged and sold.

Oysters are filter feeders, which means that they absorb their food by pumping water through their systems and extracting the small organisms they need. If they pump polluted water, some of this may remain in the oyster when it is harvested. Oysters are eaten either raw or lightly cooked and the whole animal is eaten, except for the shell. If an oyster is harvested from polluted water and then eaten raw or cooked only lightly so that bacteria in the meat are not destroyed, there is a real risk that harmful organisms from the water can be passed to the oyster consumer, potentially causing a variety of illnesses. Oysters can also concentrate marine biotoxins, known as red tides, which are potentially dangerous to consumers though not to the oysters.

To protect consumers from these risks, which apply also to mussels, clams and whole scallops, there is a federal control system called the National Shellfish Sanitation Program (NSSP) which covers all aspects of shellfish growing and harvesting and requires states to inspect and approve waters from which shellfish are harvested. The NSSP also requires all shellfish containers to be tagged or labeled with codes indicating the original packer or shipper. All shellfish users must keep records, including the tags, so that in the event of a problem, any affected shellfish can be traced and, if necessary, removed from the market. (See NSSP for further information.)

Oysters are available live or as shucked meats. Increasing quantities of oysters are sold live, in the shell. This is partly because there is increasing interest from restaurants in serving them raw, on the half shell; partly because labor costs in producing areas are rising and it has become easier to transfer the shucking task to the end user. Oysters in the shell are also used for preparing cooked products such as oysters Rockefeller, which are presented on the half shell.

Shellstock is usually sold by volume measures and sometimes by weight. Most states define volume measures to provide some consistency. Size grading is arbitrary, with regional and even individual preferences. Terms such as standards or selects mean different things to different packers: find out what is meant before you buy. Table 5 shows a common size grading for eastern oysters. Shellstock may be offered as individual oysters (singles) or as clumps, which should be cheaper since they are harder to handle and may include some dead shell. Live oysters are usually marketed under the name of the bay or area where they were grown. Westcott and Appalachicola oysters are not distinct species, simply geographical origins.

Table 5. Common size grading for eastern oysters.

Standards	about 200 to 250 per bushel
Selects	about 100 to 200 per bushel
Large, counts or extra selects	fewer than 100 per bushel

Handling shellstock depends partly on where it comes from. All seafood should be kept cold, but oysters from warm water, such as the Gulf of Mexico, should not be cooled too suddenly or the shock will weaken them. The NSSP requires a maximum temperature of 45°F. Many shippers will keep them much colder than that. Do not let live oysters sit in meltwater from ice, as fresh water kills them. Melting ice must be given ample room to drain. However, it is important to use ice to keep the atmosphere damp. Oysters die if they dehydrate and they dehydrate if they are open. Keeping them cool and damp encourages them to stay closed. Some shippers pack them tightly into sacks so that it is harder for them to open. Oysters seem to survive better if they are shipped and stored with the deeper shell downwards. This is hard to achieve in sacks. Some growers are packing their oysters in layers in shallow cartons. This protects them much better against knocks and ensures that they can be kept the right way up. Do not use oysters with cracked shells, because they may have been contaminated with bacteria. Do not use dead oysters under any circumstances. If one is open, tap it gently. If it remains open, it is dead and must be discarded.

Frozen shellstock and half shell product is increasingly available. There are numerous half shell products and preparations offered. Because most bacteria are killed if kept frozen for several days, some safety experts recommend using frozen product instead of live. Oysters respond extremely well to freezing. It is very difficult indeed to tell if an oyster has been frozen. It is surprising that more users are not insisting on frozen product.

Most oysters are used as meats, which are available fresh and frozen in a variety of forms. The liquor that comes with fresh meats should be semi-translucent, not opaque, although sometimes the liquor may thicken if too much heat was used to shuck the oysters. Meat color is not an indication of freshness since it varies from area to area. Smell is the most reliable indicator of freshness. Oysters always come with their liquor, but should not have added water. Many buyers specify that there should not be more than 15 percent liquid in the container. Determining whether water was added to a pack is not easy, since oysters seem to give off and then reabsorb water during transportation. If you need to check the drained weight of oyster meats, use Method 18.013-15 from

117

the Association of Official Analytical Chemists. There is a version of this test in the Code of Federal Regulations, as follows:

> The oysters are drained on a strainer or skimmer which has an area not less than 300 square inches per gallon of oysters, and has perforations at least 0.25 inches in diameter and not more than 1.25 inches apart. The oysters are distributed evenly over the draining surface of the skimmer and drained for not less than five minutes.

The AOAC method requires draining for exactly two minutes and is a better standard to use. Buy from suppliers you trust and remember that some liquid is an integral part of the product.

Some water may be added to the pack by a technique called blowing, which is designed to remove grit and mud from the meats. Air is bubbled through tanks holding oyster meats. This can induce the meats to absorb water, especially if salt is added. Federal rules limit blowing to 30 minutes (15 minutes if salt is used).

Federal regulations stipulate size grading for oyster meats from both the east and west coasts. Pacific oysters are generally much larger than eastern, so the size grades are quite different. However, few packers pay attention to the grading definitions, using instead their own definitions. Consistency of size, which is also defined in the Regulations, is often as important as the actual meat count. Make sure you know what you are supposed to be buying.

Oyster meats are usually frozen in molds, making them convenient for breading and frying. However, too much glaze on frozen meats can be a problem. Frozen meats are graded (counts per pound) and packed IQF or block. They are convenient and generally inexpensive item.

Oysters are farmed in various way. On the east coast, oystermen reseed the grounds and keep them clean of predators. On the west coast, as in many other countries, oysters are grown on rafts or ropes suspended in the water. Oyster aquaculture is well developed and hatcheries supplying seed are technologically strong. The newer technique of remote setting, which means shipping oyster larvae to be placed into water and set as oysters at the grower's location, has opened opportunities for growing oysters in many more areas.

Although oyster production in the traditional areas of the east and Gulf coasts has been hit hard by diseases, pollution and natural disasters, it is recovering. West coast production has been increasing slowly, but there are excellent prospects for increasing output from many Pacific regions. Oyster hatcheries working with the Pacific species have successfully developed triploid oysters, which do not spawn and so do not lose condition after spawning. These oysters can be harvested and eaten all year. Traditionally, oysters were not eaten in summer months (those lacking an "R"), partly because distribution was too difficult in days before refrigeration and partly because ripe and spent oysters are considered less palatable. Production is strong in Asia and although it is very difficult to secure the necessary approvals under the NSSP to import oysters, supplies worldwide seem to be in good shape. As more Americans discover the pleasures of eating oysters, the product to supply them should be available, provided that growing water can be kept clean. Pollution has caused growing areas in many states to be closed. Cleaning up the environment is an important priority for oyster growers.

The following paragraphs describe the major oyster species available to U.S. buyers.

AMERICAN OYSTER: *Crassostrea virginica*. Other names are the eastern oyster, Atlantic oyster and cove oyster. This is a cupped oyster with an elongated shell, marketed at about three inches long and growing to about twice that size. It was once found from the St. Lawrence to Mexico, but stocks have diminished dramatically in the northern part of the range. In recent years, production from the Chesapeake, which was once full of oysters, has been sadly reduced by diseases, pollution and natural disasters. Nevertheless, eastern oysters from the Atlantic and Gulf coasts still provide the greatest part of domestic production.

PACIFIC OYSTER: *Crassostrea gigas*. This is sometimes called the Japanese oyster because the species was introduced to the west coast from Japan. It is also a cupped oyster. It is fast growing and reaches over 12 inches, though it is usually harvested by the time it has grown to six inches. Pacific oysters are grown on the west coast of the United States and Canada in comparatively small quantities, though the industry is thriving and expanding. Thanks to new rules, Alaskan growers are beginning to establish oyster farms. Worldwide, this robust oyster represents three quarters of total production. It is grown in large quantities in Japan, Korea, Spain, France and many other countries. It has thrived almost everywhere it has been introduced.

Although Pacific oysters are larger, growers are learning to harvest them sooner so that they can be sold in competition with eastern oysters. The shells of both species are similar. The meats differ in that the Pacific oyster has a dark frill, while that on the American oyster is pale. These minor differences are not significant in most circumstances. Flavor, texture and meat content of oysters varies enormously form river to river and month to month. Generalizations are often made but seldom valid.

Kumamoto oysters are a small, deeply cupped strain of Pacific oysters that has found favor with buyers. Hatcheries have also developed triploid Pacific oysters, which are genetically manipulated so that they do not spawn. This means that they are palatable year round. Unfortunately, it has so far proved impossible to produce a batch that is 100 percent triploid and there is no way to tell the sterile oysters from the regular ones until they are opened. This is hampering marketing efforts.

FLAT OYSTER: *Ostrea edulis*. The European flat oyster has almost disappeared from its eastern Atlantic range, although it was an important product for at least 2,000 years. It is a small oyster with an almost circular shape and a shallow cup. Small numbers are grown in Maine and California. These are sold live for restaurant use and they are expensive.

OLYMPIA OYSTER: *Ostrea lurida*. This is also called the native oyster and the western oyster. It is a small, flat oyster, usually less than three inches across when fully grown. It takes as many as 2,000 meats to fill one gallon. It used to be enormously abundant on the West Coast, but pollution and overfishing have destroyed the resource. Olympia oysters are now a gourmet curiosity.

OTHER OYSTERS: There are a large number of species worldwide, but because of the necessary restrictions of the NSSP, few countries have been approved to ship oysters to the United States. Live oysters are occasionally seen from Chile. These are *Ostrea chilensis*, a flat oyster which is sometimes offered on menus as a Chiloe oyster, named after the region in Chile where it grows. Because these are at their best during the southern hemisphere's winter and our summer, they have established a niche in the market.

The suminoe oyster, *Crassostrea rivularis* is, like the Pacific oyster, native to Japan. It is being grown experimentally in Oregon, apparently with excellent results.

P

PACKAGING Good packaging serves two major functions. One is to protect the product during distribution so that shelf life is as long as possible. The other is to display the product so that it appeals to the end user.

Fresh fish is highly perishable. It becomes inedible very quickly, changing condition and quality as it deteriorates. Newly caught, properly handled fish is totally different from stale, ten day old fish. Many items also lose weight as they age, through evaporation and drip of moisture and soluble proteins. Anything which helps to extend the period in which fish remains in top condition is of great value to the industry. Fresh fish also lacks eye appeal to the consumer. Poets seldom sing of beautiful fillets. Many negative attributes such as "slimy" and "smelly" are associated with fish. Ideas which help to make fish attractive to the consumer are always needed. The best packaging can go some way towards overcoming both of these problems.

Vacuum packing of fresh foods, including smoked products, is now common. Provided that the film used is oxygen permeable, it should be a safe process. However, it does very little to extend shelf life. There are developments of vacuum packing which, in the long term, may offer great benefits.

In the United Kingdom, modified atmosphere packaging (MAP) is used. This looks attractive and extends shelflife. The air is removed from the package and replaced with other gases, such as carbon dioxide. The idea is to retard the growth of spoilage bacteria and extend shelf life. The process is not at present permitted in the United States. It is feared that the gases used in MAP may mask off odors, so that consumers would lose the warning that product was bad. There is also a risk that the pack could turn anaerobic and permit the growth of botulinus bacteria, especially if the package is not maintained at consistently cold temperatures, close to freezing point. The success of the packaging in the United Kingdom has depended heavily on the fact that the major user controls its distribution chain from packing plant to retail store. Without the guarantee of temperature stability, the risks of the process increase. The use of controlled atmosphere packing (CAP), a similar process for bulk shipment, is possible.

Sous-vide products are refrigerated packs that are prepared by cooking the food after the bag has been vacuum sealed. These are distributed under refrigeration and heated for the table by boiling the bag and its contents. As with MAP, it is vital to maintain the temperature of the product very close to the freezing point. The process kills the spoilage bacteria that make food smell bad (one FDA official says "Putrid smell is God's early warning system that food has turned rotten") but may leave much more dangerous organisms which multiply if the temperature is allowed to rise.

PADDLEFISH *Polyodon spathula*. Sometimes incorrectly called spoonbill catfish. This is a large freshwater fish found in the central part of the United States. Fish of 30 to 40 pounds are usual and 90 pound fish are taken. It has few bones. Where paddlefish is available commercially, it is sold as chunks of skinless, boneless hot smoked meat. The roe is sometimes used to produce a caviar substitute.

PAN FISH Any small fish prepared for the pan by removing the head, guts and gills, scraping off the scales and trimming off the tail and fins. Many small fish, especially southern species, are prepared and sold in this way, fresh and frozen. Croaker, blue runner, butterfish, small flounder that have insufficient flesh to justify filleting, scup and whiting are among many that are often presented in pan ready form. The end user can simply flour and fry the fish without any further preparation.

The common American aversion to having bones served with fish reduces the appeal of many pan fish species. However, many cooks insist that flavor is enhanced if fish is cooked on the bones. Most pan fish are cheap and can be attractively presented in restaurants.

PARASITES Tapeworms, flukes and roundworms that may be found living in the flesh of many species of fish. Some fish and shellfish are more likely than others to contain parasites. Calico scallops are prone to parasites. Swordfish may have enormous worms growing in their sides. Cod from certain areas has small worms in the muscle. Black sea bass is often riddled with parasites.

There are four things to keep in mind about parasites:

1. Firstly, they are destroyed and usually made invisible by cooking. Dead parasites are not harmful to the consumer. The objections are aesthetic.

2. Secondly, the increasing use of raw fish in sushi and similar forms requires much greater attention to be paid to parasites. Certain worms may pass alive into humans eating raw seafood and become parasites within the person.

3. Thirdly, candling is required by any good packing house. Candling means holding each fillet in front of a light so that any parasites can be seen and then removed. Cod, flounder and similar groundfish are routinely candled. The process is quick and inexpensive and avoids much grief. Candling also reveals if any pinbones have been left in product intended to be boneless.

4. Finally, the existence of parasites in a particular fish resource is frequently transitory and may indicate that the fish is not being heavily used. Wild fish stocks tend to expand to the limit of their food supply. When industrial fishermen start to deplete the numbers, the remaining fish have the same food supply to share between them, so each fish can get more to eat. This improves the health of the individual fish, which become less vulnerable to attacks of parasites. Well known examples of this are Labrador cod, which was heavily infested when first properly exploited, and Peruvian whiting, which also harbored plenty of parasites. Both species are now clean. The effects of heavy fishing enabled the remaining fish to improve their health and to resist the parasites. However, environmental changes can alter matters, too. The growth in the numbers of seal, now that they are protected animals, has increased the incidence of parasites in cod and other fish. This is because certain parasites have a complex life cycle that includes spending a period in seal droppings. More seals provide more habitat for the worms, which then infest the cod.

The lesson from this is that you should not make assumptions about which fish may have parasites, since it is likely to change if fishing continues. Also, parasites probably relate to small populations of fish. Labrador cod was infested while Nova Scotia cod was not: do not condemn a species on the basis of experience with fish from one area.

The health risk from parasites is small and could be largely eliminated by freezing fish that is to be eaten raw. Fish held at 0°F for a week will be free of parasites. In Japan, much sushi is made from frozen fish. In the United States a number of large chains only serve sushi only from fish that has been frozen.

PASTEURIZING The application of sufficient heat to kill most bacteria, but not enough heat to cook the product. Killing the bacteria prolongs shelf life, since some of the bacteria cause spoilage. Blue crab meat is commonly pasteurized, giving it an extended shelf life under refrigeration (see CRABS). The alternative, freezing, harms the texture of this and certain other products, so pasteurizing is a preferred process.

Pasteurizing is not an easy technique. Careful control and monitoring are required and anything less than spotlessly clean conditions can reintroduce bacteria and make the whole thing pointless.

PERIWINKLE

Littorina littorea. This is a small sea snail, about one inch long, found in large numbers along the Atlantic shoreline as far south as Delaware Bay. The name is also sometimes used for whelks in the United States. In the United Kingdom, periwinkles are commonly called winkles. Small quantities from Maine are sold live in New York, but there is only a small market there and virtually no market elsewhere. Live periwinkles (never use dead ones) are boiled for about ten minutes in salted water. They can be extracted from the shell with a pin. Apart from the operculum (which is the flap which seals the shell closed) all the soft parts inside the shell are eaten.

PIKE

There are numerous names covering a fairly small assortment of species, all from northern freshwater lakes. They are good quality and high priced fish, which vary little in use.

WALLEYE: *Stizostedion vitreum.* Walleye pike is the other approved name. Blue pike, doré, pike-perch and yellow pickerel may be used but are not officially sanctioned. Walleye is not actually a pike but a close relative of the lake perch, which it resembles. It is commercially important throughout the Great Lakes and the central lakes of Canada. Walleye and yellow pike are probably the most common names.

The flesh is excellent, sweet, finely flaked and very white. The meat is lean and has a good shelf life. It is versatile and can be used pretty much the same way as sole; that is to say that almost any cooking method will work well. Commercial fish are generally between one and three pounds.

Product is available dressed or as fillets. Large quantities are imported from Canada, fresh and frozen. Like many freshwater species, walleye is particularly popular in the midwest. Its use in gefilte fish recipes gives it appeal in many other areas also.

SAUGER: *Stizostedion canadense.* This is a smaller version of the walleye, similar in use and taste. It should only be called sauger, according to FDA rules. You may find it called sand pike, yellow walleye and, again, doré.

PIKE: *Esox lucius.* A widely distributed game fish between five and 10 pounds. Commercial supplies come mainly from Canada, in the form of dressed fish, fillets, skinless fillets and steaks. The eating qualities are good, though it is considered not as good as the walleye. Pike which have not been commercially processed should be scalded before use to remove the slime and loosen the scales.

MUSKELLUNGE: *Esox masquinongy.* This species is very similar to the pike, but grows much larger, often over 10 pounds and reaching 30 pounds.

PIKEPERCH: Both walleye and sauger may occasionally be called by this name. The term is not approved by the FDA for any species.

PICKEREL: *Esox americanus* is the grass pickerel. *Esox niger* is the chain pickerel. These are small fish, seldom growing over 12 inches. They are popular gamefish, similar in eating quality to pike.

PINBONES
A row of small bones running horizontally along the lateral line of many fish, from the nape backwards for about one third to one half the length of the fish. In small fish, the pinbones soften sufficiently in cooking so that they are not noticeable. The pinbones of slightly larger fish, such as herring, are softened in pickling and smoking. In very large fish, such as chinook salmon, the pinbones are large enough to be easily removed when the fish is cooked.

Pinbones can be removed by cutting a V-shaped piece from the fillet. Since this removes meat from the thickest part of the fillet, it reduces the yield and so is expensive. In some species, the pinbones are avoided by cutting the fillet diagonally to remove pinbones and belly flap together.

Fillets without pinbones are described as boneless fillets. Fresh fillets are rarely shipped boneless, however. Some fish, such as flounder and sole, do not have pinbones at all, so the fillets (if properly trimmed) are boneless.

Bones are a considerable marketing problem for fish products. Although consumers in more fish-oriented parts of the United States such as the northeast and northwest, may have the experience and knowledge to cope with pinbones on their plates, many consumers throughout the country avoid fish because of the possibility of bones. Anyone who has ever suffered the pain of a small bone in the gum will sympathize.

POLLOCK
Pollachius virens. Among the many illegal names applied to this fish are Boston bluefish, blue cod and green cod. Saithe, coley and coalfish are terms used in the United Kingdom for the closely related eastern Atlantic species, *Pollachius pollachius*. The FDA permits the name for this species to be spelled pollack, in accordance with British usage.

Pollock is related to cod and similar fish, with darker flesh and more pronounced flavor than either cod or haddock. It is invariably cheaper than cod. Pollock grow to about 35 pounds, but four to ten pound fish are usual commercially. Fresh pollock is handled exactly like cod and used in much the same way. It is most readily available in late fall, when it is spawning, but some supplies are offered throughout the year. Domestic landings in the northeast are supplemented with imports of fresh and frozen product from Canada.

Pollock is oilier and darker colored than cod. The flesh is grayish white when cooked. It has a large flake. The flavor is stronger than cod. Pollock is an excellent fish and offers good value for the money. The higher oil content, however, reduces the shelf life and it does not have the extended keeping qualities of cod. Most product is sold as skinless fillets. There are also small regional markets for dressed or headless and dressed fish for steaking.

The name Boston bluefish appears to have originated in Canada. It is thoroughly confusing, since the fish is quite unlike bluefish and more expensive. It is wiser to avoid using this name, which is not recognized by the FDA.

ALASKA POLLOCK: *Theragra chalcogrammus.* Legally, this may be called pollock, Alaska pollock or walleye pollock. Marketers also try to call it by more enticing names, such as snow cod and snow scrod. These flights of fancy have been shot down by the FDA.

Alaska pollock meat is whiter and leaner than that from the Atlantic fish, much closer to cod in appearance. The flesh is softer and the flake smaller than cod. It is one of the major groundfish resources of the world and huge quantities are used for fish blocks and surimi. The roe is prized in Japan and Taiwan. Frozen skinless and boneless fillets are widely available. Fresh pollock is less common, as most of the fishery uses large factory trawlers that process and freeze the catch at sea.

POMFRET
Brama brama (Atlantic pomfret) and *Brama japonica* (Pacific pomfret). Sea breams, shaped like large butterfish. The Atlantic species is caught occasionally as a by-catch. The Pacific species is reportedly found in the open ocean throughout the Pacific in large quantities, especially during the summer. It has good quality meat, similar to pompano. Since there is little or no U.S. high seas gillnetting, this fish is not caught by American vessels. The species is a potential major supply resource.

POMPANO
Trachinotus spp. The nomenclature for pompano is confused. All of the following may be legally called pompano, but they sell for sometimes substantially different prices under the alternative names.

POMPANO: *Trachinotus carolinus*. A deep bodied, compressed Atlantic fish (like a butterfish), with silver skin changing to golden on the belly, pompano sells for extremely high prices. The flesh is white and delicate. Larger fish may be filleted. Smaller fish are usually offered in pan ready form. In price terms, however, it is definitely not just another pan fish. Most pompano run between one and two pounds. They are caught from the Carolinas, around Florida and into the Gulf. The main fishing season is October through May.

Pompano are excellent broiled, but there is a substantial tradition of cooking them in pouches by baking them and using complex sauces. The delicacy and quality of the flesh, however, should be able to stand on its own with simple preparation and cooking. Whether pompano is worth the sometimes exotic price charged for it has to be determined by the end user.

PERMIT: *Trachinotus falcatus*. This is remarkably similar to the pompano and is also from the Atlantic. The only external difference is in the number of rays on the dorsal fin. The permit's second dorsal fin has one spine and 17 to 21 soft rays; the pompano's second dorsal fin has one shorter spine and 22 to 27 soft rays. (Rays are the rib-like rods in the fin.) To the layman, the permit has a longer dorsal fin than the pompano. The chances of finding fish in commercial distribution with the fins sufficiently intact for positive identification are poor. Permit are found in the same areas at the same time as pompano.

There are two major differences between the pompano and the permit. Firstly, the permit grows much larger. Pompano seldom exceed five pounds. Permit may reach 50 pounds. Fish weighing 20 to 30 pounds are often taken by anglers and commercially caught fish around 10 pounds are also common. The second major difference is that pompano generally costs much more.

In terms of eating quality, permit is said to be drier and coarser than pompano. However, pompano also gets less desirable as it gets larger.

Since the visible difference between the two species is so slight, the probable commercial explanation is that smaller fish are offered as pompano and larger fish as permit. When buying these species look for smaller fish and try not to pay pompano prices for permit.

PALOMETA: *Trachinotus goodei*. This is also very similar to the pompano and permit. It is slightly different in appearance, as it has four dark gray bars vertically on each side and the big dorsal fin is considerably longer than on the pompano or permit—sometimes so long that it extends behind the tail when flattened against the fish. The palometa's body is slightly more compressed laterally, so the fish is not so good for filleting as it provides a thinner fillet. In practice, dealers seem to call this fish a permit, also.

PORGY *Stenotomus chrysops.* This fish may also be called scup. It is a sea bream. Do not confuse porgy with pogy, which is menhaden.

Porgies are abundant off the middle Atlantic coast some years. Huge quantities used to be trapped in Narragansett Bay. The fish is generally small (about one pound on average) and is bony. A large part of the catch is exported and frozen product which does not succeed in finding an export market can be a glut on the domestic market.

Fresh scup can be used as a panfish. Unfortunately, there are many small bones. Larger fish are easier to eat, because the bones can be seen more easily, but larger fish have a coarser and less pleasing texture than the smaller ones. The scales are strong and the skin tough. It is important to scale the fish soon after it is caught. Scup can also be hot smoked.

PRAWN Shrimp. The definition of a prawn depends on where you are. The FDA permits the word to be used on the label of larger canned shrimp, provided that the word shrimp appears as well. In the United Kingdom, prawn usually means a large shrimp (although the word also applies to a lobsterette or langoustine). In Australia, tropical shrimp are called prawns. In the Pacific northwest, the locally caught northern shrimp are often described as prawns. These are much smaller than the widely available tropical shrimp, such as those from the Gulf of Mexico. Freshwater shrimp are sometimes described as prawns.

It is better to ignore the word prawn and use shrimp exclusively.

PUFFER *Sphoeroides spp.* These fish can be toxic. They are usually marketed in the United States as sea squab, although the FDA does not accept this name. Other (unapproved) names include blowfish and globefish, but the best known name is the Japanese word fugu.

The organs and viscera of puffers can be deadly. Proper preparation is supposed to ensure that toxins do not contaminate the meat. Nevertheless, a number of Japanese die each year from eating fugu, even though it is prepared in special restaurants by specially trained and licensed chefs. The FDA has been unable to prevent the Japanese puffer from being imported.

The Atlantic northern puffer is supposed to be only mildly toxic. It is offered cleaned and skinned, usually as butterfly fillets labeled sea squab.

R

RANCIDITY Oxidation of fats. Fish become unpalatable when rancid. The problem is worse with oily fish than with lean fish. Salmon, sablefish, herring, tuna and other oily fish must be stored carefully. It is mainly because of the oxidation of the fat that oily fish have a generally shorter shelf life than lean fish.

Rancidity is easy to detect. Even frozen fish, if rancid, gives off a recognizable and unpleasant odor. Belly cavities produce the most rancid odors very quickly. Rancid fat is also easily seen by the yellowing patches on cut surfaces, especially belly flaps.

Freezing fish will slow down rancidity, but not prevent it. If frozen storage temperatures fluctuate, even though remaining well below freezing point, frozen fish will turn rancid very quickly.

RATPACKING Ratpacking is the practice of putting poor quality product underneath a layer of good quality material. If you inspect an open container, you will see what seems to be tolerable fish. If, however, you take the time and trouble to probe through the container, you may find product which is not acceptable.

RECALL The FDA can enforce the removal from distribution of any product linked with an outbreak of food poisoning or of product which is found to be in violation of food and drug laws. Where seafood is concerned, recalls primarily apply to frozen product, since fresh product causing trouble is unlikely to be around by the time the link between the product and the problem is established.

However, molluscan shellfish subject to the controls of the Shellfish Sanitation Program may be recalled if any cases of illness or contamination are traced to particular batches or areas. It is partly for this reason that the system requires distributors to keep records of where the shellfish comes from and where it is sent.

There are three classes of recall. Class One is the most serious, involving possible death or serious illness or injury. Class Three recalls cover such things as misbranding. Any recall is expensive. The better your records of where your product has gone, the easier and less costly a recall becomes. Insurers will sometimes cover the identifiable costs of a recall.

REDFISH
Sciaenops ocellata. Do not confuse this redfish with the North Atlantic species called redfish or ocean perch (see ROCKFISH). *S. ocellata* is a drum and may be called drum or red drum. It should not be called channel bass. Redfish is a Florida and Gulf species, presently largely reserved for recreational fishing. If available, it is used for baking and for chowders and can also be fried. It became known when blackened fish, Cajun style, became popular. Preferred size is about five pounds.

The resource seems to have declined in recent years. At one time, red drums of up to 85 pounds were regularly caught off the Carolinas, but very little is now landed from the Atlantic at all. The meat is lean, firm and white, with a large flake and moist texture. Larger fish have tougher flesh, more suitable for chowders and stews. This species may contain trematode parasite worms, so it must be cooked thoroughly. However, it is a tasty fish and extremely popular, especially on the Gulf coast.

It seems unlikely that redfish will be available in any quantity commercially for some years, although efforts are being made to develop aquaculture techniques. Should it reappear, the fish is likely to be valuable again.

RED TIDE
Red tide is the popular name for blooms of algae which, when ingested by certain shellfish, produce toxins that poison people eating the shellfish. By extension, red tide has become the generic name for these toxins (there are at least 18 different ones). The illness they cause is known as paralytic shellfish poisoning (PSP) and it can be extremely serious. The poison attacks the nervous system, affecting breathing and muscle control. In extreme cases it can be fatal. The term red tide relates to the reddish tinge in the water when some of these algae bloom. Domoic acid poisoning, which occasionally is found in mussels, is caused similarly by an algae bloom. This is now monitored together with PSP.

PSP is caused by a number of different algae which occur in many parts of the world. There appear to be more occurrences than there once were, though this may be due to increased awareness of PSP, which leads to better diagnosis and reporting. In the United States and Canada, it is present, from time to time, in most coastal waters. The toxins do not usually appear to harm the shellfish (or, at least, not very much). Oysters, mussels, clams and scallops are all affected. The toxins concentrate in different parts of different species. In scallops, the viscera are vulnerable but these are seldom eaten in the United States. In butter clams, the siphon seems to concentrate the most poison and retains it for longer than other parts of the clam.

To protect the public (and the industry) from red tide, the states operating the NSSP monitor all growing waters for the causative algae and test shellfish regularly to make sure they are free of toxins. If PSP is found, growing waters are closed until the shellfish have purged themselves and are again clean. This may take weeks or even months, depending on the outbreak. This is only one of many reasons why you must ensure that all shipments of clams, mussels, oysters and live scallops are properly tagged and labeled with the license number of the shipper. Illegally harvested shellfish can kill your customers and will do no good to your business.

RIGOR MORTIS
Stiffening of a body after death. Fish passing through rigor should not be frozen, straightened or filleted since the flakes will be torn apart. See GAPING. It is better to dress fish either before the onset of rigor or after it has passed through rigor and the muscles have relaxed. Filleting is better left until after rigor.

RIPE FISH
Ready to spawn. In general, ripe fish has comparatively poor quality flesh, since the fish's metabolism is concentrated on creating the roe or milt.

ROCKFISH
This section covers Atlantic ocean perch and Pacific rockfish. The nomenclature is confusing. Note that STRIPED BASS (*Morone saxatilis*) is known as rockfish in some areas.

OCEAN PERCH: *Sebastes marinus*. This species may also be called redfish, which is the usual name in Europe. It is caught with two other virtually identical species, *Sebastes mentella*, the deepwater redfish and *Sebastes fasciatus*, the Labrador redfish. The name redfish should not be confused with the red drum, which is also called redfish in the Gulf states (see REDFISH).

Ocean perch are slow growing, deep water North Atlantic species reaching about eight pounds but usually, nowadays, caught at well under two pounds because fishing pressure has removed most of the older and larger fish. The skin is red and the flesh slightly pink and moist, cooking to a fine white color. It sells particularly well in the midwest, as an alternative to lake perch. Domestic landings are almost entirely in Maine and Massachusetts. Large quantities are imported from eastern Canada. Iceland supplies some larger fillets.

Fillets are sold with the skin on and range in size from under one ounce to as much as one pound, though there are very few large fillets produced domestically any more. Although most perch is frozen, fresh product is available from domestic, Canadian, European and (via air freight) Icelandic processors. The red color gradually fades as perch gets older after catching, so the skin color is an indicator of the freshness of the fish.

Small fillets are breaded and often used as a substitute for lake perch. Larger fillets, especially if the skin color is good, may be displayed illegally as a substitute for red snapper. This is a perfectly good fish in its own right. The flesh is quite sweet and firm and has little fishy taste.

ROCKFISH: *Sebastes spp.* Both identification and naming of this large group of fish is complex. The FDA has attempted to simplify it, although its requirements are not always followed. In summary, these are the naming rules for Pacific rockfish:

Ocean perch: one species, *Sebastes alutus.*
Bocaccio: one species, *Sebastes paucispinis.*
Chilipepper: one species, *Sebastes goodei.*
Cowcod: one species, *Sebastes levis.*
Treefish: one species, *Sebastes serriceps.*
Rockfish: all of the above, plus all the others, a total of 63 species.
Redfish: name not permitted.
Pacific snapper: name not permitted.
Red snapper: name not permitted.
Rock cod: name not permitted.

California has state rules for rockfish sold within its borders:

Pacific ocean perch: ten species.
Pacific red snapper: 13 species.

Washington and Oregon also permit the use of Pacific red snapper as a name. Note that fish so labeled shipped between these states is in violation of federal rules, even though the name is legal in both shipping and receiving states.

Table 6 shows the California nomenclature for the 23 species of rockfish, where it differs from the federal standard. All of these fish may be called rockfish, inside and outside California. For the full list of rockfish species, see the Appendix.

Rockfish is still a large and important marine resource, even though it now supports far less fishing than it did in the 1960s. Most of the fish is trawl caught and processed on land. Some boats keep the fish in champagne systems (bubbling air through tanks of slush ice to keep it circulating), to maintain the maximum freshness and quality. Fish

Table 6. Rockfish: Sebastes species names used in California that are not permitted in interstate trade.

Common name (FDA approved)	California name Not legal for interstate trade	Scientific name
Bococcia	Snapper, Pacific red	*Sebastes paucispinis*
Chiliopepper	Snapper, Pacific red	*Sebastes goodei*
Cowcod	Snapper, Pacific red	*Sebastes levis*
Rockfish, aurora	Perch, Pacific ocean	*Sebastes aurora*
Rockfish, bank	Snapper, Pacific red	*Sebastes rufus*
Rockfish, black	Snapper, Pacific reds	*Sebastes melanops*
Rockfish, canary	Snapper, Pacific red	*Sebastes pinniger*
Rockfish, darkblotched	Perch, Pacific ocean	*Sebastes crameri*
Rockfish, olive	Snapper, Pacific red	*Sebastes serranoides*
Rockfish, redbanded	Perch, Pacific ocean	*Sebastes babcocki*
Rockfish, redstripe	Perch, Pacific ocean	*Sebastes proriger*
Rockfish, rougheye	Perch, Pacific ocean	*Sebastes aleutianus*
Rockfish, sharpchin	Perch, Pacific ocean	*Sebastes zacentrus*
Rockfish, shortbelly	Snapper, Pacific red	*Sebastes jordani*
Rockfish, shortraker	Perch, Pacific ocean	*Sebastes borealis*
Rockfish, speckled	Snapper, Pacific red	*Sebastes ovalis*
Rockfish, splitnose	Perch, Pacific ocean	*Sebastes diploproa*
Rockfish, stripetail	Perch, Pacific ocean	*Sebastes saxicola*
Rockfish, vermillion	Snapper, Pacific red	*Sebastes miniatus*
Rockfish, widow	Snapper, Pacific red	*Sebastes entomelas*
Rockfish, yelloweye	Snapper, Pacific red	*Sebastes rubberimus*
Rockfish, yellowmouth	Perch, Pacific ocean	*Sebastes reedi*
Rockfish, yellowtail	Snapper, Pacific red	*Sebastes flavidus*

Sources: FDA *Fish List* and *Menu and Advertising Guidelines for California Restaurants*, Sea Grant.

caught on long lines may be alive when brought onto the deck. Live fish can be bled, which improves shelf life as well as appearance. Line caught fish is a superior product, especially for the fresh market.

Some rockfish have brown or greenish skin, some have red skin. All have firm flesh, bland flavor and good shelf life. Some long line caught fish is sold dressed, but the greatest part of the rockfish catch is filleted. Producers like the red skinned fish because of its "snapper" connotations and lighter colored flesh. Generally, red skinned rockfish are sold with the skin left on the fillets. These may be sold as red snapper. Brown skinned fish are usually sold skinless, perhaps labeled rock cod. Most rockfish are quite small, up to about three pounds, giving fillets from as small as two ounces up to about ten ounces. For use as perch in the midwest and elsewhere, the small fillets are preferred since they more closely resemble the freshwater perch that consumers may think they are buying.

The freshness of rockfish is easily determined by the appearance of the skin, which should be shiny and bright. Wrinkled skin that looks too large for the fish is a sign of stale product. There is a growing market for these fish alive, to be shipped to restaurants and retailers where they are displayed and sold from tanks.

All rockfish species grow slowly and the resource is fished hard, making future supplies unpredictable. Seasonally, the fall spawning season brings the largest quantities and the lowest prices. Huge quantities of these fish were caught in the past and conservation measures seem to be slowly restoring some of the populations.

ROCK SHRIMP

Sicyonia brevirostris. A hard shelled shrimp which has a flavor and texture a little closer to lobster. Rock shrimp have become firmly placed on the market since machinery was developed to peel them. Rock shrimp are common off the Atlantic coast of Florida and throughout the Gulf of Mexico. Landings are highest in the late summer and early fall, but some rock shrimp is landed year round.

Most rock shrimp is about 40/50 count and yield from the tails is only about 50 percent. Consequently, they sell for substantially less than regular shrimp. Peeled rock shrimp can be an excellent buy. However since it has a tendency to shrink, it must not be overcooked or you lose the advantage of the lower initial price. Cooking for 20 seconds should be sufficient. The texture also toughens considerably if the shrimp are overcooked. This warning applies to all forms of rock shrimp. Cooking times should be much less than for regular shrimp, whether you are boiling, broiling, baking or frying the product.

Shell-on rock shrimp is quite difficult to handle because of the hard and spiny shell. They can be bought split, which helps considerably. Split rock shrimp broil very well. As a retail item, rock shrimp have appeal. The product is inexpensive, for good tasting shrimp. The low yield and difficulty in peeling should be explained to the consumer.

Fresh rock shrimp have a very short shelf life. Speedy and careful handling is essential. In good condition, fresh or frozen raw rock shrimp have somewhat transparent flesh and a mild odor. Discoloration of the flesh and development of strong odors indicate the product is stale. In general, it is preferable to handle rock shrimp in frozen form rather than fresh, unless you are close to the source of supply and are sure of adequate sales to move purchases as soon as they arrive. Raw tails lose quality within 24 hours. Cooked rock shrimp will last several days under refrigeration.

ROE Fish eggs; female gonads. When about to be laid, the roe of many fish is as much as a third of the body weight of the fish and certain fish roes are extremely valuable. Although few are eaten in the United States, those that are tend to be expensive. Some roes are also exported to countries which appreciate these delicacies; herring, mullet and capelin exports, which are sometimes substantial, are largely based on Asian appetites for the roes of these fish.

By contrast, many consumers in the United States regard roes as inedible. The half moon shaped roes on scallops, for example, are almost always removed. Traces of scallop roe in frozen packages are regarded as reason for rejection. Yet scallop roes are a delicacy in Europe, where scallops are preferred if they have the roes attached. All roes have a very high fat content and must be moved and consumed rapidly.

Some roes that are available in the United States include:

STURGEON roe, which is the most valuable. See CAVIAR.
COD roe, often smoked, is valued throughout Europe but can only be obtained here from a few specialty suppliers.
HERRING roe, which is the basis for substantial industries on both the Atlantic and Pacific coasts that supply either herring with roe, or the roe itself, to Asian buyers. It is hardly ever eaten in the United States. European markets buy large quantities of Atlantic herring roes.
MULLET roe, similarly exported as whole fish or as roe, and mainly sold to Taiwan and other Chinese markets, is also not consumed here.
POLLOCK roe, from the Alaska pollock, is another Japanese delicacy which has yet to appeal to American tastes.
SALMON roe has been sold to the Japanese for many years and is made into red caviar, which is sometimes eaten in the United States. Some salmon eggs are also used for making bait for sport fishermen.
SEA URCHIN roe is a very valuable commodity. The product is difficult to handle, but overseas markets are not hard to find. There are considerable resources of urchins in the United States.
SHAD roe is actually eaten in the United States. Normally sold in natural pairs (that is, intact, as they come from the fish), shad roe is an accepted delicacy in certain areas, mainly along the Atlantic coast. It is important to handle the roe gently, so that the sac is not broken.

Small quantities of roe may be used as garnishes or in sauces. Mullet, for example, which is an inexpensive fish, can be upgraded to gourmet status by garnishing with some of its roe. The spread of sushi restaurants has increased demand for many different roes, including those from flying fish and capelin.

S

SABLEFISH *Anoplopoma fimbria.* This species may be called black cod in California. It is sometimes mislabeled butterfish, apparently because of the shiny, buttery appearance of the fillet along the nape. Despite the FDA's preference, the fish is generally known as black cod, even though it is not related to codfish.

Sablefish is a Pacific species caught the whole length of the North American coast from Alaska to California. Harvested fish average about eight pounds and range between two and 12 pounds, though the maximum size of the species is about 50 pounds. The skin is dark gray to black and has a distinctive furry texture. The flesh is oily, containing about 15 percent fat. When cooked, the color is pearly white. It has large flakes and a flavor close to that of salmon. It is one of the finer fish to eat and is underrated on the U.S. market. It is best prepared using recipes that work well for salmon. Barbecuing and broiling give excellent results. Most sablefish is sold smoked, in which form it is popular in many parts of the United States especially in New York and other major Eastern cities, where it is a staple item of many delicatessen departments.

Sablefish is caught in trawls, on longlines and in pots. Potted and line caught fish is traditionally considered better because it was bled and because it was handled more carefully. Product is now available frozen at sea, which is also very good quality. Fish from trawlers that bring the fish ashore in ice is less likely to be of such good quality. However, it is Japanese demand that drives this market, so all fishing is done with that in mind. Unfortunately, if fish is not good enough for Japan, it is unloaded onto the domestic market. Sablefish is generally available as headless, dressed fish or as skinless fillets. As with most oily fish, it does not have a particularly long shelf life.

Most of the U.S. supply is exported, chiefly to Japan, in frozen, dressed form. Japanese buyers, who sell the fish for sashimi, prefer fish over five pounds dressed weight and are very fussy about quality.

Smokers in the United States also prefer such larger fish. As a result, there is often a surplus of cheap, smaller fish which can be filleted and offered at very reasonable prices.

Sablefish is beginning to be appreciated on the west coast, where kasu cod has become popular. This is a Japanese preparation where the fish is marinated in the lees from sake brewing.

Supplies seem reasonably assured. On the basis of limited evidence, various regulations have been imposed, limiting catches and restricting the minimum size of fish landed to 22 inches. However, it seems that there are ample sablefish available. If these supplies were supported by more marketing efforts to educate users about the clear value of sable as a fresh fillet, likely increases in demand could be met. Some early research is being done in Canada on the possibilities of farming sablefish.

Sablefish is an excellent item which deserves more attention from domestic marketers.

SALMON

Salmo salar is the Atlantic salmon. There are six Pacific species, as follows:

Oncorhynchus tschawytscha is chinook salmon, also called king and spring.
Oncorhynchus keta is chum salmon, also called keta.
Oncorhynchus kisutch is coho salmon, also called silver and, when canned, medium red salmon.
Oncorhynchus gorbuscha is pink salmon, also called humpback.
Oncorhynchus nerka is sockeye salmon, also called red and blueback.
Oncorhynchus masou is cherry salmon.

All of these Pacific species, except cherry salmon, are found in U.S. waters. Cherry salmon is found only on the Asian (western) side of the Pacific. In addition, steelhead, *Oncorhynchus mykiss*, an anadromous form of rainbow trout, looks, behaves and tastes very much like Atlantic salmon, although it lives in Pacific waters. In market terms, it is effectively salmon and is therefore treated in this section as a salmon.

All wild salmon come from the Northern hemisphere. With the rapid growth and success of salmon farming technology, Atlantic salmon is now grown in the southern Pacific (Chile, for example) and Pacific cohos are farmed in the Atlantic (France, for example). Because of the growth of salmon farming, fresh salmon is now available year round almost everywhere in the United States, at affordable prices. Salmon remains a highly regarded fish by consumers and gourmets alike. It is no longer an unusual treat, but is fast becoming a staple of the high quality seafood department.

The life cycle of the salmon is well known. Salmon are anadromous. Atlantic salmon is capable of spawning, returning to the sea and then going back again to spawn in its stream. All the Pacific species spawn once and then die in their streams. Most salmon are caught on their

137

way to or up their river when they are intending to spawn. The fish stop feeding when they reach fresh water and use their stored energy (fat) to fuel their considerable efforts to swim upstream to the spawning grounds. Consequently, the condition of the fish deteriorates as they get further from the ocean. Salmon caught in the ocean will have fat flesh, shiny silver skins and be in the best condition for eating. The same fish, if taken hundreds of miles upstream, will have watermarked skin, pale, lean flesh and a watery texture that will be almost inedible. In between, there are many different grades which buyers have to differentiate.

Salmon are the basis for substantial recreational fisheries, and in the Pacific northwest, certain Indian tribes are given important rights over harvesting the fish. Federal, state, provincial and even internationally negotiated conservation rules govern catching salmon: when and where they may be caught, with what sort of gear, who may fish for them, how many may be caught and so on. Salmon, especially on the west coast, is one of the most closely regulated fisheries. The regulations are complex and are designed to protect and expand the resource. Consequently, there are times when the regulations seem to have been framed to make marketing as difficult as possible: in some years, for example, almost all supplies will cease just before a major holiday, because conservation rules have closed a number of important fisheries. Despite the irritations, there can be no doubt that the overall effect of the conservation efforts has been strongly positive. There is plenty of salmon and supplies for the future seem well assured. The stocks of salmon, especially in Alaska, have recovered from the overfishing and pollution that reduced the catches in the 1940s and 1950s. Without the rules, salmon were disappearing fast from the west coast, just as they had almost completely gone from the east coast. Now, the Pacific species are well managed and the application of similar principles to the Atlantic salmon is beginning to restore quantities to many rivers where the fish had been either rare or extinct. Even so, there is virtually no legal commercial harvest of wild Atlantic salmon. Quantities permitted in the United Kingdom, Ireland and Canada are minute compared with the huge business in farmed salmon and wild Pacific salmon.

Farmed salmon, of course, is not subject to conservation regulation. It can be grown wherever suitable environmental and regulatory conditions coincide. Norway pioneered the domestication of Atlantic salmon and now dominates markets. Scotland, Canada and Chile are major producers. There are at least 10 other countries farming salmon commercially, with more planning to enter the business. While Alaska has perhaps the best unexploited environment for salmon farming, the state has decided to ban finfish aquaculture permanently, because of fears that the farmed fish will compete with the wild harvest on which so many Alaskans depend. The farmed fish clearly already competes and will no doubt be grown elsewhere. Contrast the Alaskan attitude with that in Japan, where fishermen's cooperatives have entered the salmon farming business in order to be able to supply their customers throughout the year.

Producers and marketers of farmed and wild salmon in the United States have campaigned against each other, claiming that their own product is superior. The efforts each has made to decry the other have harmed both the perpetrators and the seafood business as a whole: consumers do not know, and probably do not care, whether their fish was farmed or hunted. They care that it should be fresh and wholesome, handled in a clean manner and available at a reasonable price. Both farmed and wild salmon can offer these benefits. Consumers hear the cries of battle and leave the battlefield. Such disputes may benefit chicken producers. It is hard to see how they benefit the seafood business.

There are certain differences between wild and farmed salmon, though these are neither generic nor, in most instances, sufficiently pronounced for anyone but an expert to be able to identify them. Farmed fish is more likely to be bled. Bleeding improves the color, shelf life and flavor of the flesh. Troll caught salmon is gutted on the fishing boat and so is well bled. Net caught salmon is mostly iced and landed at shore plants for processing and so is not bled. On the other hand, farmed fish may be highly stressed before slaughter, especially if they are transported distances in live tanks. Stress reduces the glycogen in the flesh and causes some loss of flavor. Wild fish is often maturing and the flesh quality deteriorates as it swims up the river. Farmed fish is unlikely to be maturing (though it certainly may occur). Overall, the fish that is best handled and processed is likely to be the best one to eat. This depends on so many variable circumstances that generalizing between farmed and wild fish is a waste of time. Note that taste tests on wild and farmed fish carried out under the sometimes biased auspices of trade groups have produced results that are consistent with random selection theories.

ATLANTIC SALMON: *Salmo salar.* This fish is seldom called anything but salmon or Atlantic salmon. You may see it with geographical names such as Norwegian salmon. Commercial supplies are now imported from as far away as Chile, so the Atlantic designation is somewhat confusing: many Atlantic salmon are now grown in the Pacific Ocean. At one time, salmon were found throughout the northern Atlantic. They spawned from the Connecticut River to north of Labrador on the western Atlantic; from Russia's White Sea south to Portugal in the east. They were particularly abundant in eastern Canada, the British Isles and Scandinavia. Now, there are very few wild salmon anywhere in the Atlantic.

Atlantic salmon will grow to 100 pounds. Farmed fish is usually harvested when it is between six and 12 pounds. The fish is distinguished from other salmon by the cross shaped black markings on the silver skin.

Total world commercial supplies of wild salmon, mainly from northern Canada, are probably less than 3,000 metric tons annually (though in the United Kingdom and Ireland, poachers increase the supply beyond recorded levels). Strenuous and expensive efforts are being made internationally to restore salmon stocks, but it is unlikely that these efforts will ever produce enough fish for commercial exploitation.

World production of Atlantic salmon (from fish farms) exceeded 220,000 metric tons in 1989. Norway was the major producer. Scotland, British Columbia and Chile are all important; there are at least 14 countries that claim production on a commercial scale. The United States produces only tiny quantities, on both coasts. Alaska's recent decision to ban farming of marine fish will ensure that U.S. producers never have a dominant position in the farmed salmon business.

Atlantic salmon is mostly sold fresh, dressed. Air shipment from overseas suppliers reaches most parts of the United States very quickly. It is feasible to fly as little as 500 pounds almost anywhere. The flesh is pink and moist. It looks attractive in a retail display as dressed fish or as steaks or fillets. Size grading depends on the producer, but most farmers grade in small steps, so that sizes of the fish in a box are very close. Farmed fish ranges from as little as two pounds to 10 or 12 pounds. Fish under about four pounds are often harvested because the farmer needs some cash and cannot wait for the fish to grow larger. This has been a particular factor from western Canada. Fish around six to nine pounds are regarded as prime sizes. Smaller fish often have paler flesh; the color develops and is retained better in older fish. Norwegian and Chilean fish are quality graded and inspected to national standards.

Wild salmon harvests are intensely seasonal, as most fish are caught in the short period as they congregate before entering their spawning streams. Fish farmers can extend their harvest seasons almost year round, so supplies of fresh product (if you prefer that to frozen) can be reasonably assured. In practice, salmon farmers try to harvest their fish when wild fish is scarcer. Pacific species are readily available in the summer months, so farmed Atlantics dominate the market in the winter and spring, when the wild fish cannot be supplied in fresh form in any significant quantity.

Because of high continued levels of production, more Atlantic salmon is being frozen. Norway is even experimenting with canned products (although these could not be sold in the United States since the federal Standard of Identity for canned salmon requires the product to be made from one of the Pacific species, not from Atlantic salmon). Frozen salmon is easier and more convenient to use. Because it does not have to be shipped by air, it is often less expensive. Norwegian producers are leading the way in developing value added products (both fresh and frozen). These include skinless and boneless fillets, salmon cakes and a variety of steak and fillet products. Such products, combined with good packaging, could help develop new markets to absorb the increasing supplies.

Despite reports that Norway is scaling back future production plans, more farms are coming on line in many other countries. Atlantic salmon, which is so far the easiest salmon species to domesticate, is likely to remain plentiful.

STEELHEAD: *Oncorhynchus mykiss*. This is the anadromous strain of the freshwater rainbow trout. Note the scientific name, which was changed from *Salmo gairdneri* in 1989. *Salmo irrideus* is another outdated scientific name for this species. It is called steelhead or steelhead trout in the United States and steelhead salmon in Canada. The fish is almost indistinguishable from Atlantic salmon in size, flesh color and taste. Smoked steelhead is a little less red than similarly smoked salmon. The skin has the characteristic rainbow hues and speckling of the trout. It is farmed in increasing quantities, especially in Norway and Japan. New strains, developed in the United States but little used here, provide red fleshed fish with deep bodies that grow exceptionally fast. The fish are normally under 12 pounds, though much larger fish have been caught by anglers. The natural range of the steelhead is from California to the Aleutians. There is virtually no commercial fishing for steelhead in the United States, where the species is reserved for Indian and recreational fishermen. Commercial supplies are available from Indian enterprises as well as from Canada.

Because steelhead is generally cheaper than Atlantic salmon of equivalent quality and size, it can be a very good buy, when it is available.

CHINOOK: *Oncorhynchus tschawytscha*. King or spring salmon are permitted names. Tyee and blackmouth are officially discouraged. Tullie or tully refers to poor quality, mature fish. Chinooks are caught from California northwards through Alaska, as well as in Asia. They run at different times in different areas and are therefore available fresh, in small quantities, through a large part of the year. Because many of the fish are caught after they enter rivers, peak supplies of chinooks are in June, July and August. Chinooks enter the Copper River and Prince William Sound in the middle of May. In May and June, fish from California should be marketed. Also in June, they are caught in western Alaska, primarily the Yukon. In the later part of June, there are troll fish available from southeast Alaska and in July these should be available from the Washington and Oregon coasts. Substantial quantities of chinooks run up the Columbia system in September. However, since openings and closings on individual rivers are very specific and fisheries are closed for periods at the height of the salmon runs to allow spawning fish to escape through to the spawning grounds, all of this is only the most approximate guide to availability. Fortunately, chinooks can be farmed and quantities are available year round, mainly from Canadian and domestic growers.

Chinooks are the largest Pacific salmon, growing to 30 pounds and more (actually to over 100 pounds in a few reported instances). Four pounds is about the smallest that may be taken legally. Chinooks are graded under-7, 7/11, 11/18 and 18 pounds and up, but there are substantial variations between shippers in chinook grading. Some may simply distinguish between small, medium and large, in which case the buyer requires more information.

The flesh is red and the skin brightly silver: chinooks are most attractive in appearance. They are also excellent fish for smoking, especially the troll caught fish. As with all salmon, the skin color develops watermarks and the flesh color fades as the fish swims up the river to the spawning grounds. The lowest quality category is sometimes called a tullie. This has mottled skin and greyish flesh. It is soft and has a limited, low priced market. In some places these are heavily smoked to give flavor and to dry the flesh to improve the texture, but the result is far from being good smoked salmon.

Pale kings are red kings which have started to lose their color. That is to say, they are somewhere between a red king and a tullie. White chinooks have white flesh, naturally. These are a separate race of the species. Although the flavor is similar to red chinooks, market preference for red meated salmon restricts sales to areas near the fishing grounds, where people are familiar with the product.

COHO: *Oncorhynchus kisutch*. Also legally called silver salmon and medium red salmon. Other names include jack salmon, which are young, mature males; silversides; and blueback (which is incorrect) when canned. It is better to use the term coho than silver, since silver can more easily be confused with silverbright, which is a grade of chum salmon.

Cohos are found throughout the whole range of Pacific salmon. There are also substantial numbers of landlocked cohos in the Great Lakes, though these are very inferior in eating quality. Cohos average six to 12 pounds when caught. The largest fish grow over 30 pounds. The flesh is lighter in color and texture than that of king salmon. Troll cohos are almost as well liked by smokers as troll kings. Grading is usually 2/4, 4/6, 6/9 and 9/12 pounds. Coho skin and flesh deteriorates when the fish enters fresh water. Red skinned silvers are an inferior grade, although the flesh color is usually fairly pink, so the fish may be used for steaking.

Farmed cohos are widely available from growers in Chile, Canada and the United States In addition, small farmed cohos are being sold at sizes under one pound, dressed and prepared much like fresh trout. These baby salmon are delicate and similar in flavor and texture to trout.

PINK: *Oncorhynchus gorbuscha*. The species may also be called humpback. It is caught from Oregon to Alaska and is also the most abundant salmon on the Asian side of the Pacific. Exploitation of the resource in Siberian rivers is increasing. Pink salmon runs tend to be heavy in alternate years and the fish is usually good value fresh or frozen in the years of high availability. Pinks are not farmed, probably because their market value is low.

Most of the fish are netted in the estuaries. Because the fish generally spawn near to the sea without traveling far up the rivers, watermarking is seldom a problem. Pinks develop a humped back as they move towards spawning and this feature will indicate fish from fresh water. Pink salmon spoil and turn rancid faster than other salmon species. Troll fish, if available, is superior, probably because it has been bled.

CHUM: *Oncorhynchus keta*. The official names for this species are chum and keta, though the latter is seldom if ever used commercially. Terms such as silverbright, semibright, calico and dark refer to quality grades and not to the species. Do not confuse silverbright chums with silver salmon (coho).

Chums are caught from Washington northwards to the Bering Sea and down the Asian coast to Japan, where the species is most highly regarded. Chums average 10 to 12 pounds when caught. Fresh fish should be graded 4/6, 6/9 and 9/12 pounds. Fresh chums should be available from June until November or December, with peak supplies in July and August. Some rivers have two runs of chums and in these the summer run is generally small fish and the fall run much larger fish.

Chums from the ocean have bright, silver skins and red meat. Most packers describe these as silverbrights, although purists may insist that the true silverbright can only come from the Johnstone Straits between Vancouver Island and the mainland. As the fish moves into fresh water, the skin develops watermarks and the flesh loses some of its color. Semibrights are chums where the watermarks do not extend below the lateral line on each side of the fish. Semibrights should still have good meat color and are suitable for steaking, where the skin appearance is less important. Fish described simply as bright chums or brights will usually be semibrights; otherwise, the seller will use the full term silverbright to make sure he describes his fish at the best possible market value.

Chums watermarked below the lateral line will probably be called simply chums or may be fall chums or calicos. Eventually, the skin turns dark grey with mottling and the flesh loses almost all its red or pink tint. The fish is then called a dark chum, which is the lowest quality

SALMON

designation. These fish are cheap. They can be used for steaking but have little flavor. Most of this fish is frozen, not sold fresh.

Chums collect numerous other names, many of them local. The important characteristics are the flesh color and the skin color. It is safer to ask for fair descriptions of these than to rely on individual interpretations of the names.

SOCKEYE: *Oncorhynchus nerka*. The species may also be called red and blueback, though both of these names are generally used on canned salmon rather than fresh or frozen. Quinault is a race of small sockeye on Washington's Olympic Peninsula. Kokanee is a dwarf landlocked sockeye. Sockeye is an abundant species in major river systems in Alaska and British Columbia, where it is the most valuable species for the canned salmon industry. It is caught in small quantities in Washington. Attempts to farm the species have not yet been successful, although hatchery techniques are well developed. Alaska's abundant recent catches have been greatly enhanced by hatching sockeye and releasing them in the streams.

Sockeyes have deep red flesh and bright silver skin. They range up to eight pounds and may be graded 3/5 and 5/8 pounds. Although sockeye is the prime salmon for canning, its flesh color gives it great popularity for eating fresh also. It is highly regarded by Japanese buyers, who import large quantities of fresh and frozen sockeye from U.S. and Canadian processors.

Sockeye is available fresh from March to July and then again in August and September from the Fraser River. Generally, when it is available, there are large quantities in a short time.

CHERRY: *Oncorhynchus masou*. This species, sometimes called masu salmon, is not native to North America, but has a restricted range in Japan and Korea. It is similar to a small coho. Chilean salmon farmers are experimenting with the species. If they are successful, then the fish is likely to appear on the market in the same forms as farmed coho salmon.

SALMON TROUT: This term is illegal in the United States. It is used in Europe for seagoing rainbow trout (that is to say steelhead). It is sometimes used for steelhead, for which it seems a reasonable name. However, some growers of freshwater rainbow trout have also used it, which seems less appropriate. The FDA says that it "does not recognize the name salmon trout for any fish sold in interstate trade in the U.S.A."

OTHER SALMON: There are no other fish species which may be called salmon in the United States, or in Canada. Species from such places as

Argentina and Australia, though they may be called salmon locally, are not related.

COLOR: Each species of salmon has certain characteristics that affect its particular market. Overall, flesh color is probably the most important of these characteristics. Japanese consumers buy a large proportion of the Alaskan sockeye catch (there are few sockeyes in Japan and the species so far has not been amenable to farming) because of its bright red flesh. Salmon smokers in Europe favor the Atlantic salmon and the chinook because of their deep pink color. Salmon with good color will generally sell for a higher price than salmon with less color. The simplest way to define the color of salmon flesh is to use the *Pacific Salmon Color Guide* from BC Research, 3650 Wesbrook Mall, Vancouver, BC, V6S 2L2.

HANDLING FRESH SALMON: Fresh salmon should be well iced (or properly chilled and kept refrigerated with gel packs). When you receive salmon, fill the belly cavity with ice and bury the whole fish in more ice. Do not steak or fillet it for use until you need it. The additional cut surfaces will hasten dehydration and rancidity. Pick the fish up by the head, not the tail, to reduce the risk of tearing the backbone and damaging the meat.

If you receive farmed salmon with rather soft flesh, this may indicate that it was over stressed when harvested. Excessive scale loss is another symptom of poor handling. Stressed fish uses its glycogen reserves so the meat will have less flavor and sweetness.

Quality standards for salmon vary. There is no U.S. Grade standard for whole or dressed fish (a draft one was dropped many years ago), though there is one for salmon steaks. If you need standards for designing your own specifications, use the Canadian government standard or that from the Fisheries Council of BC. For the intrinsic quality and handling of farmed salmon, Norway and Chile both have good standards.

How salmon is harvested, whether from the wild or from a pen, affects its quality and shelf life. Other factors impacting quality include processing, packing and distribution. The whole subject of salmon is examined in much greater detail in this author's *Salmon—The Illustrated Handbook for Commercial Users*, published by Van Nostrand Reinhold/Osprey Books.

SALTED CURED AND DRIED FISH
Stockfish, bacalao, salt fish, salt cod, and so on. Salt draws moisture from the flesh, making the flesh less attractive to bacteria and thus preserving it. All of these terms apply to fish that has been preserved by salting or by salting and drying or by salting and smoking. Modern cures are less

SALTED CURED AND DRIED FISH

concerned with preservation, since refrigeration does the job better with far less damage to the texture of the meat. Instead, curing has become a flavoring process to create palatable and attractive products. However, traditional cured products are still made, especially in Canada, although markets are mainly overseas. See SMOKED FISH for further information.

STOCKFISH: Headless, dressed, hard dried, unsalted fish, usually cod. Atlantic pollock, ling and hake may also be used. Stockfish needs no refrigeration and is widely used in tropical countries, especially Africa and the Caribbean. At times it is a major product utilizing cod that would otherwise be sold fresh or frozen. Pacific cod as well as Atlantic cod is now used for stockfish production.

SALT COD: Bacalao (bacalhau in Portuguese) is split and stacked in layers with large amounts of salt for several weeks, by which time it is cured, so it will keep indefinitely without refrigeration. The salt removes moisture from the flesh and replaces the moisture partially with salt. The low moisture and the high salt combine to make an environment hostile to bacteria growth.

Salt cod was the product which lured fishermen from Spain and Portugal to the Grand Banks not long after Columbus sailed. It is still a very important product and traditional methods remain in some places. For a marvelous description of life aboard a Spanish pair trawler fishing for cod and salting it, see *Distant Water* by William Warner (published by Atlantic-Little Brown).

Salt cod is now produced in the Pacific as well as the Atlantic. It is also marketed in milder forms which require some refrigeration. Boneless salt cod is popular in the northeast, consisting of pieces of salt cod, more or less without bones, often sold in wooden boxes containing one pound. The boxes possibly attract some purchasers more than the contents.

The price of cured fish depends partly on the type of cure and partly on the moisture content of the product. Many, probably all, cures are traditional products based in particular regions and often seasonal. Since the fish flesh varies at different times of the year, so does the cured product made from it. Traditional cured fish was made from cod, with an occasional use of ling, haddock and similar cod-like species.

Some of the cures are briefly defined as follows.

AMARELO: Portuguese cure. Salt content about 18 percent. Yellow color.

BRANCO: Portuguese cure. Salt content about 20 percent. White color.

DRY SALTING: See kench, below.

FALL: Light salt, pickle cure, 45 to 48 percent moisture.

GASPE: Light salt, pickle cure, 34 to 36 percent moisture.

HARD: Dry salted, dried to moisture content 40 percent or less.

HEAVY SALTED FISH: Has about 40 percent salt (dry weight basis) and moisture content with a range of definitions, in Canada, as follows:

Extra hard dried: less than 35 percent moisture
Hard dried: up to 40 percent
Dry: 40 to 42 percent
Semi dry: 42 to 44 percent
Ordinary cure: 44 to 50 percent
Soft dried: 50 to 54 percent

Heavy salted soft cure fish will have about 17 percent salt (dry weight basis) and about 47 percent moisture.

KENCH: Dry salting: Split fish and salt in alternate layers, allowing the moisture to drain. This is a basic curing method.

LABRADOR: Heavy salt (about 18 percent); 42 to 50 percent moisture.

PICKLE CURE: Wet salting: the fish are salted in a container so that they are cured in the liquid, or pickle, that forms.

SHORE: Light salt, kench cure. About 12 percent salt and 32 to 36 percent moisture.

SOFT: Dry salted, dried to over 40 percent moisture.

Stockfish is unlikely to be requested in the United States Handling it is easy. Cooking it is more of a problem as it must be soaked for days before use. Salt fish should normally be kept chilled, in spite of any label advice that ambient temperatures are acceptable.

A great deal of world supplies of cod are used for stockfish and salt cod. The markets for both types of product are highly volatile and very much dependent on the abilities of the consuming countries, which are mostly rather poor, to buy and pay for their needs. Consequently, every so often, producers may switch production to frozen cod products, which as a result become more available and less expensive.

SANDLANCE *Ammodytes spp.*

A small fish, up to about five inches in length, resembling eel in appearance (so sometimes called sandeel), which can be caught along many beaches in great numbers.

SARDINE

Sandlance are the food of many important commercial fish but are not much used in the United States. They can be used effectively for whitebait.

SARDINE In the United States, sardine is a generic term describing canned fish of various herring like species. Small quantities of frozen and fresh sardines are imported from Spain and Portugal. These may be whole or dressed, but if they are not frozen, dressed product is preferable. Sardines should be graded by length. Smaller sizes, under seven inches, are considered superior. The flesh is fat but light and delicate.

PACIFIC SARDINE: *Sardinops sagax*. Also known as the pilchard or California sardine, this fish has largely disappeared, although quantities were seen during the period when the weather disturbances collectively known as El Nino were changing the pattern of Pacific marine life. The species has been replaced in the ecosystem by the anchovy. The sardine is now a protected species and can normally only be caught as a by-catch with mackerel.

SPANISH SARDINE: *Sardinella anchovia* and *Sardinella aurita*. These small fish are found on both sides of the Atlantic. They are frequently used for bait in Florida, where they are apparently available in large quantities, although few are caught because there is no developed market for sardines as food. This is a pity as they are good and very cheap.

If sardines should be available, they are best when small (under seven inches) and need to be very fresh as they develop stale tastes quickly. Skins should be shiny, and scales should be removed by thorough washing. They should be unmarked, without tears or cuts.

The fish are headed and gutted and used for broiling, frying or barbecues. The flavor is distinctive and very good. It is important to have sardines with a fairly high oil content. If they are low in fat, they taste dry.

SASHIMI Raw fish sliced thinly and eaten. This is a Japanese word for a Japanese seafood preparation technique. It has become highly popular in the United States in recent years and is spreading beyond Japanese restaurants.

Sushi, although rather different in Japan, is a word used in the United States to describe essentially the same thing. The distinction generally made in the United States is that sashimi is the sliced raw fish, served with soy sauce and horseradish dips; sushi is the more elaborate preparation using dried seaweed (nori) and rice as well as other decorative ingredients. The preparation and serving of the many sushi dishes are an art form requiring great skill and lengthy training. See also SUSHI.

It is obviously essential that the fish served be extremely fresh and in every way in top condition. Not all seafoods are considered suitable for eating raw. Herring, mackerel, salmon and bluefin tuna are among the most highly regarded varieties. Squid, abalone and other shellfish are also used. So are flounders, bass and many other seafoods.

Eating fish raw involves the risk of absorbing parasites, especially tapeworm and roundworm. Freshwater and anadromous fish are particularly likely to contain parasites which are harmful to consumers. Most, probably all, of these parasites can be destroyed by freezing the fish. Frozen fish, provided it was in top condition when frozen and is handled correctly by the freezer, is entirely satisfactory for sashimi and sushi. If you serve raw fish, it is safer to serve fish that has been frozen.

SCALES Almost all fish are covered in one of the four types of scales. The commercial significance of scales rests in two areas. Firstly, only fish that have true scales are permitted for Kosher diets. Secondly, the scales of most fish have to be removed before the fish is used since they are small, hard and often sharp and cannot be eaten.

Fortunately, scales of most smaller fish are very small and are removed by the usual washing processes. Some larger fish have scales which must be scraped off. Most fish is scaled by the first receiver. All fillets are scaled, since the scales are taken off before the fish is cut. The scales of fish such as salmon are shiny and reflective. They give the fish its bright sheen, which is so valuable when it is displayed in a retail case. However attractive they are, scales must be removed before the fish is cooked or you risk them falling into the meat. Some scales can be as unpleasant as fish bones if eaten.

SCALLOPS Filter feeding molluscs with two flat shells. In the United States, the adductor muscle is the only part normally eaten. This is the cylindrical white meat which holds the two shells together. In Europe, the semicircular orange and brown roe is eaten with the adductor muscle, at least from larger scallops. In Japan, all the soft parts inside the shell are eaten, both raw and cooked. In the United States, small numbers of scallops are now eaten whole, steamed. Because of the risk of red tide toxins from the viscera, whole scallops have recently been added to the National Shellfish Sanitation Program, which means that shipments must be tagged to prove that they come only from licensed dealers. Receivers must keep the tags for at least three months. See SHELLFISH SANITATION.

Scallop farming is an important industry in Japan, but is so far not successful on a commercial scale in the United States and Canada. France, Spain and China are other countries where scallop aquaculture appears to have developed further than it has in North America.

The U.S. scallop market is basically divided into markets for three

SCALLOPS

types of scallops. The market as a whole is dominated by sea scallops, harvested in the United States and Canada from adjacent areas of the northwest Atlantic. To supplement supplies, frozen scallop meats are imported from Japan and other countries. The much smaller bay scallop sells for a premium price but is usually scarce. Domestic production is sold fresh, but imported frozen scallops, such as queen scallops from Europe, are thawed and sold as substitutes. Calico scallops, which are smaller and cheaper than bays, are often sold as substitutes for bay scallops. Calicos and other small scallops from Mexico and other countries to the south are imported in both fresh and frozen states. Many of these are sold as bays rather than as calicos. The more important scallop species offered fresh to U.S. buyers are discussed below.

SEA SCALLOP: *Placopecten magellanicus*. This is the correct scientific name preferred by the American Fisheries Society. *Pecten magellanicus* is an older and no longer correct designation. This scallop is landed in Canada and the United States. New Bedford, Massachusetts, is the major U.S. center. The auction here plays a large part in determining the price of fresh (and of frozen) sea scallops. Sea scallops are caught by draggers on the offshore banks. Because scallops do not live long out of water, they are shucked by hand on the boats so that only the meats are brought ashore.

In the last few years, controls on the size of scallops have been tightened. To oversimplify the complex rules, meats counting smaller than 30 per pound may not now be landed. The idea is to protect the breeding potential of this heavily fished resource by allowing scallops to grow to a larger size, and therefore through reproductive cycles, before they may be caught. Canadian and U.S. regulations are roughly aligned. Product sizes may run from under 10 per pound—which is very rare—to whatever the changing legal minimum may be.

Cloth bags containing about 30 pounds each are the traditional method of shipping sea scallops. Bags may be packed three to an ice filled wooden fish box. Problems with weights of scallops, which absorb water and drain themselves readily, are discussed below.

Many sea scallops are size graded and frozen. The graded product has advantages and frozen product is easier to handle than fresh. If sea scallops are scarce or expensive, Japanese scallops (*Patinopecten yessoensis*), may be imported as an alternative. These have rather wider and flatter meats than the sea scallop. The species has been cultured experimentally in British Columbia. There is a possibility of farmed supplies from this area in the future.

BAY SCALLOP: *Argopecten irradians*. These are a smaller, inshore scallop, found from Cape Cod to North Carolina. This true bay scallop is also known as Cape Cod scallop or Cape scallop, Long Island scallop,

SCALLOPS

Peconic scallop (Peconic Bay in eastern Long Island being a major source area) and Rhode Island scallop.

It is widely considered the best flavored and nicest scallop to eat. Fresh bays, if really fresh, are excellent to eat raw and require very little cooking at any time. Because of their excellent quality, scarcity and high price many inferior scallops are passed off as bay scallops. Taste, however, is a simple and reliable test for the genuine article.

Bay scallops are caught inshore and fishing is restricted to hand dredging, in most places with strict limits on daily catches. These conservation methods have kept the fishery operating in much the same way it always did, with fishermen limited in both the seasons they may take the animal and the amount they can bring ashore, so it is still almost an artisanal fishery. "Brown tide" algae blooms have disrupted the fishery in recent years. The algae destroys many of the scallops, in ways that are not yet clear.

Bay scallops are landed alive in the shell and shucked by hand on shore, usually by the fishermen. The meats are packed into plastic gallon containers. There are various interpretations of the weight of meats in a gallon. In practice, the weight varies between seven and nine pounds. Volume measures for product that is to be resold by weight are unsatisfactory, but this practice is unlikely to change and the buyer should both know his supplier and check the weights and volumes of the shipments received. Bay scallops are not graded. They tend to be fairly uniform in size from any given shipment. Counts range from 50/60 per pound down to about 100 per pound. The normal range is around 70/90 count.

Bay scallops are very expensive. Seasons are also limited. Fishing starts in September and although in most places is allowed until the following April, most fishermen find the catches unrewarding after the first two or three months, so landings usually drop to a trickle by the turn of the year and the onset of the coldest weather.

Small quantities of bay scallops are sold live for eating whole. Scallops sold in this way come under the jurisdiction of the NSSP and must be tagged to show they come from licensed dealers. It is likely that all scallops will be brought into the program in time.

Samples of bay scallops sent to China in the early 1980s appear to have naturalized successfully. Frozen Chinese product has been offered on the U.S. market and is expected to increase in volume. It is possible that the species is mixed with a native scallop, *Chlamys farreri*.

Because of their scarcity and value, thawed imported scallops are popular substitutes for bays. At least, they are popular with the dealers who sell them. The species most favored for this purpose is the queen scallop, *Chlamys opercularis,* from northern Europe. Although this an excellent scallop, note that a thawed product may have a shorter shelf life than a refrigerated one.

SCALLOPS

Calico scallops (see next paragraph) are sometimes labeled as bay scallops, especially in retail outlets. The price and taste differences between the two are significantly large. Calicos are thinner and longer than bays. They are also whiter and less creamy.

CALICO SCALLOP: *Argopecten gibbus*. This is a smaller and less delicious species, found from the Carolinas down to Brazil and landed fresh in large quantities from Florida and the Carolinas. Smaller quantities are landed from the Gulf coast also. It is a semitropical scallop and grows very quickly, which means that the beds, although frequently reported to have been wiped out by overfishing, restore themselves very quickly. Supplies of calico scallops vary greatly from year to year.

Calicos are normally shucked on shore, but unlike other domestic scallops they are usually shucked by steam, not by hand. The equipment subjects each scallop to a short blast of steam which kills it and causes the shell to fall open. The animal is then shucked from the shell and the edible muscle removed. The steam process and the small size of the scallop may mean that the scallop meat is slightly cooked in the shucking process. This may toughen it slightly. Because the scallops are brought ashore in the shell and the areas where they are found tend to have warm temperatures, freshness also sometimes suffers.

Calicos are smaller than other domestic scallops, ranging from about 60 meats per pound down to 200 or more. Some packers are beginning to grade calicos, since there can be a substantial premium paid for the larger sizes (lower counts). Calico scallops are sold fresh in gallon containers, usually plastic pails.

PINK SCALLOPS AND SPINY SCALLOPS: Pink scallops are *Chlamys rubida*. Spiny scallops are *Chlamys hastata*. These two species of small scallop are harvested and sold together. They are found from Alaska to California. Buyers probably do not realize that they are getting two different species, although the shells are not alike; the spiny scallop, as its name suggests, has spiny ridges on the shell. The meats are very small. Economically, these scallops need to be sold whole, or there is not enough product to justify the effort of fishing for them. The two species are sometimes described as singing scallops, which is a fanciful notion of uncertain provenance.

PACIFIC CALICO SCALLOP: *Argopecten circularis*. This is better known commercially as the Mexican scallop or Panama scallop, since they mostly come from these countries. Growers on the west coast of Mexico and in the Gulf of California are producing increasing quantities of this species. Meats are shipped fresh into California and distributed along the west coast. The scallop is similar in shape and size to the Atlantic calico scallop, but is sometimes mislabeled as a bay scallop.

WEATHERVANE SCALLOP: *Patinopecten caurinus.* This is also called the Alaska scallop and the giant Pacific scallop. The meats of this large scallop are very similar to sea scallop meats. Although there are thought to be large resources of the species in Alaska, they are not fished because they share habitat with juvenile king crab, which is a more valuable resource. Dredging for scallops would damage the young crabs.

OTHER SCALLOPS: There are over 350 species around the world. A number of them enter the U.S. market, although these are generally frozen. Any frozen scallop may be thawed and mislabeled as fresh. For more detail about other scallop species and all aspects of the scallop business, see this author's *Shellfish—A Guide to Oysters, Mussels, Scallops, Clams and Similar Products for the Commercial User* (Van Nostrand Reinhold/Osprey Books). Note also that there are persistent rumors of scallops being cut from skate wings or cod cheeks. There is no credibility in such reports. Apart from the cost and labor of making such imitations, the textures of scallops, skate and cod are quite different.

HANDLING SCALLOPS: Perhaps the greatest problem associated with handling and using scallops, fresh or frozen, is the common practice of soaking them in phosphate dips. This is described and discussed under the heading DIPS. So far as scallops are concerned, the muscle is capable of absorbing plenty of water. The addition of phosphates helps the meats to hold this water. The chemical also helps to whiten the scallop, generally regarded as a marketing improvement.

The use of phosphate dips or other additives must be noted on the label.

Sea scallops packed in cloth bags, and bay and calico scallops packed in plastic pails should all be well buried in flake or chipped ice at all times. Bags of sea scallops leak into the ice, which should be changed every day or so to reduce bacteria build up. Plastic containers often appear to be well covered in ice, but cavities melt around the containers, leaving air pockets which can permit the product to warm up. This increases spoilage significantly. It is therefore important to remove the ice and re-ice the scallops. Only by doing this regularly and thoroughly can you ensure that the product is preserved as well as possible.

Scallops in their shipping containers build up a strong odor which can be rather unpleasant even when the scallops are perfectly fresh. If you are unsure, rinse a few scallops and smell them again. However, as a regular practice, scallops should never be rinsed until they are about to be used, as rinsing washes away soluble proteins and reduces the weight of the product.

Scallops that were warm when packed can sometimes be bad at the center of their containers because it takes too long to chill the middle of so large a pack. Scallops from the middle of a bag or pail should be

SCALLOPS

checked for freshness if you have any doubts— these may well turn bad before the ones on top. This applies only to scallops where the packer has not ensured that the product is properly chilled before he put them into containers. The problem is probably more likely with calico scallops, which can be warm when shucked on shore using steam.

All scallops should look and smell fresh. The odor should be sweet and seaweedy. Any traces of sourness, gas or iodine indicate spoilage. All scallops should look clear and shiny. Sea scallops are slightly creamy, bay scallops are definitely creamy. Calicos are whiter and less shiny. Traces of pink color usually indicate that a roe was cut and are *not* a problem, unless you think that customers might reject the scallops when displayed. (However, there is no reason for them to do so, except customary aesthetics.) Sea scallops should be packed rather loosely in their bags.

The whole muscle should be completely removed from the shell. Hand shucking done inexpertly tends to cut through the meat rather than around it. This can leave the scallop rather oddly shaped and also encourages it to fall apart during cooking. Scallops must never be cut along the grain (that is, vertically through the two circular surfaces that are the top and bottom of the scallop). Any vertical cut destroys the homogeneity of the meat and it will fall apart when cooked. Small pieces of what appears to be gristle on one side of the meat are actually part of the muscle and perfectly edible. These skirts should normally be removed from larger meats as a minor aesthetic defect. The meats should be cleanly separated from the rest of the animal: no roes or other organs or viscera. Seaweed, shells and other adulterants, of course, should not be present.

Scallop counts should be accurate and are easily checked by removing pieces and then counting and weighing the remainder. To check uniformity, weigh five pounds and discard any pieces (many buyers will accept only very few pieces in a shipment). Then select and weigh the largest 15 percent and the smallest 15 percent from the remainder. Divide the weight of the largest scallops by the weight of the smallest. If the answer is less than 1.5, the scallops are very uniform. If the result is greater than three, the scallops are not uniform. Note that these specifications are more rigorous than those required by the U.S.D.C. for Grade A certification, but they are closer to the current needs of the market.

Scallops are delicate. They require very little cooking and demand careful handling. Overcooking is a common offense, which detracts from the taste and texture and also harms the cook because an overcooked scallop shrinks a lot, so you must use more to give a reasonable looking portion. Many recipes suggest cooking scallops for far too long. Almost any method for two to three minutes is ample. Broiling is especially detrimental to scallops: less than a minute on each side is required if the shellfish is not to toughen and shrink.

SCAMPI
Strictly, this is an Italian word describing the LOBSTER-ETTE or langoustine. It has become common to describe shrimp scampi as a dish using shrimp in a garlic and butter preparation. Clearly, the meaning is imprecise and so the word should not be used in trade.

SCIENTIFIC NAMES
These provide a universal language for scientists throughout the world for giving clear and unambiguous identification of the species they are discussing. The system was codified internationally in 1901 and works very well. Each species has (generally) a two word name. The first is the genus, the group of similar organisms to which the species belongs. The second word is the specific name. The two together provide unique nomenclature for the species. Common practice is that if the first word is repeated in a text, then it is abbreviated to the first letter. For example, if we had just mentioned *Oncorhynchus kisutch*, coho salmon, and then wanted to refer to chum salmon, the scientific name of the chum could be written as *O. keta*. Since this practice can be confusing, in this book we normally spell out the name in full each time it is used.

Scientific names with three words are sometimes used. The third word may indicate a subspecies—an instance where the two animals are considered almost identical. Atlantic herring is *Clupea harengus harengus*. Pacific herring is *Clupea harengus pallasi*. Sometimes the third word is the name of the scientist who first described the species. In this case, the third word is not printed in italics.

Vernacular names vary between regions and, of course, between languages. The use of scientific names makes it possible to communicate clearly which species you mean. If a supplier in South America offers you sea bass, what does he mean? There is no way to tell, as sea bass can apply to hundreds of different species. If, however, he can tell you the scientific name, then you can find out exactly what that fish is and what it is normally called in English.

SCROD
Scrod is not a fish. Scrod is a size designation for certain white fleshed fish sold in the Boston fish market.

Scrod haddock is haddock between 1½ and 2½ pounds.
Scrod cod is cod between 1½ and 2½ pounds.
Scrod pollock is pollock between 1½ and 4 pounds.
Scrod cusk is cusk between 1½ and 3 pounds.

Federal agencies (National Marine Fisheries Service and the Food and Drug Administration) regard the use of scrod by itself as mislabeling, although they seldom seem to attack the practice. It is regarded in the same way as labeling a type of fish 'large' by itself, with no species qualification. Occasional actions have been taken against the use of the word for more obscure fish such as wolffish or ocean pout. On the Boston Fish Pier, the word scrod on its own normally means small haddock. The other species are designated by name, as in cod scrod which is used for necessary clarity.

There are many explanations for the origins of the word. Probably the most ridiculous is that it is an acronym for Served Catch Reeled On The Day. This theory also holds that the addition of the 'h' to make schrod indicates that the fish used is either halibut or haddock. The idea of using halibut as scrod is totally unbelievable. A more likely explanation is that the word scrod, however spelled, is derived from a Middle Dutch word, schrode, which meant slice or shred and referred to a fish too small for salting and storing so was cut up and used straight away.

Most dishes calling for scrod will work equally well with any of the four fish designated under Massachusetts law. Cod and haddock are always readily interchangeable. Without the skin, they cannot be distinguished without scientific assistance. Atlantic pollock is a little darker and oilier, with a slightly stronger flavor. Cusk is clear and white and largely indistinguishable from cod and haddock. Nevertheless, proper labeling requires that the type of fish be named, not just its size. "Baked Fresh Small" would be a peculiar description; so is "Baked Fresh Scrod."

To repeat, scrod is not a fish, it is a size designation.

Consumers, as well as many people in the seafood business, believe in the existence of scrod as a separate species. Some prefer scrod to cod and haddock. Others prefer cod to scrod. The widespread use of the term in retail markets and on menus attests to the consumer appeal of the word. When you buy scrod, make sure you know what type of fish you are getting. You should also make sure that this information is passed on to your own customers.

Remember, *scrod is not a fish*. It is a size description.

SEA CUCUMBER

These look like large caterpillars. They can be dried and exported to Asian markets. The whole animal is utilized. The only U.S. buyers of sea cucumbers are the Chinese and even this market is very limited.

SEA ROBIN

Prionotus carolinus. A small, bony fish that grows to about 12 inches. It is similar to the European gurnard and is sometimes described as gurnard. There are several species, of which the one listed here is the most common. Sea robins have wing-like fins and bony plates on their heads. They are little used in the United States The meat is firm, rather like that of monkfish. The meat yield from the individual fish is small, but the resource is thought to be substantial. Gurnards are well regarded in France and Italy.

SEATROUT

SEATROUT *Cynoscion spp.* Atlantic coast drum species. The name in Europe might mean anadromous brown trout, but that definition is not applicable in the United States.

SPOTTED SEATROUT: *Cynoscion nebulosus.* Also called speckled seatrout or speckled trout. Although speckled trout is found all the way from Virginia to Texas, most production comes from the Gulf states. Substantial quantities are also imported from Mexico, when market conditions are right. Peak catches are in the winter months in the Gulf. The recreational catch of speckled trout is many times larger than the commercial catch.

Commercially available fish are generally between one and four pounds, mostly sold dressed. Some fillets are also produced. The flesh is lean and has a small flake and a fine flavor. The fish is rightly prized. However, it does not keep well and should be handled carefully and fast. This fish may be offered simply as trout, especially in southern states.

GREY SEATROUT: *Cynoscion regalis.* Commonly called weakfish and squeteague, though the FDA does not favor these terms. Weakfish are caught from Massachusetts to Florida, but most commercial supplies are taken between New York and North Carolina. This is also an important recreational fish. As with the speckled trout, sport fishermen probably catch far more of it than do commercial fishermen.

Weakfish migrate northwards in the spring and summer, then back south and offshore (where they are seldom caught) in the fall. Supplies are generally available between April and October from different parts of the Atlantic coast. There are some landings year round from Florida.

Most fish offered are between three and seven pounds and are sold dressed. Fillets are also available, with skin on. The flesh is white and lean, with a fine, delicate flavor. Weakfish are popular along the east coast and generally the price is quite reasonable. This is often an undervalued fish, especially during the summer season, when it is readily available.

SILVER SEATROUT: *Cynoscion nothus.* Also called white seatrout. The sand seatrout, *Cynoscion arenarius* is a similar Gulf species. The two fish are probably mixed together in landings. Silver is the smallest of the seatrout, generally producing fish under 1 pounds. It is landed year round in the Gulf of Mexico, with peak landings in the winter months. Although larger fish are almost indistinguishable from speckled trout, most of the catch consists of fish well under one pound and is used for industrial purposes rather than as a food fish. Much of the catch is taken by shrimp trawlers. When sold for consumption, the small fish are generally left whole, for use as a pan fish.

SEA URCHIN *Strongylocentrotus spp.* Also called sea eggs. The roe from these echinoderms is extremely valuable, especially in Japan. Most of the U.S. fishery, which has become important in California and Maine, is targeted to this market. Sea urchins are also popular in France, but the taste has not spread to many Americans. In France, restaurants cut off the top of the spiny shell and the diner spoons out the roe. In Japan, the roes are removed and sold fresh or in various preparations used in sushi. Processed (salted) sea urchin roes and live urchins are both shipped to Japan. U.S. demand, which takes a small proportion of the catch, is almost entirely from Japanese restaurants.

Resources are easily over fished and so are much regulated, but the U.S. fishery has increased in recent years.

Roes should be brightly colored (they are usually yellow or orange) and they should smell fresh.

SEVICHE Also written ceviche. Raw seafood marinated in lemon or lime juice. The citric acid breaks down the proteins in a way similar to heating. Hot spices are often added.

SHAD *Alosa sapidissima.* Shad is an anadromous Atlantic species that was successfully introduced to the Pacific about a century ago. The roe is a considerable delicacy (see roe), but boned fish is also used, mainly fresh.

Shad are caught when entering rivers to spawn. The east coast season starts in early December in Florida and extends to May in the St. Lawrence. The run into each river is short, so buyers must find new sources each week as the fish availability moves north. On the Pacific coast, the season is later, starting in April in California and finishing in late June in British Columbia.

Female fish are generally around five pounds, males are rather smaller. The meat is white and sweet and the high oil content makes it suitable for broiling. However, the meat is also rather soft and must be handled carefully. Boning shad is regarded as a complex skill. The fish has two rows of bones, similar to pinbones, running along each fillet, one above the midline, the other below. These can be cut out as two strips (the exact location is easy enough to feel with a finger before cutting the fish).

SHARK Shark is becoming increasingly popular as a food fish, both in its own right and as an alternative to more expensive fish, such as swordfish. Many species are harvested and sold and the rate of exploitation has increased to such an extent that shark resources off the Atlantic coast, regarded as underexploited species only a decade ago, are newly subject to a management plan which sharply restricts commercial and recreational fishing. It is important to know the characteristics of the different sharks if they are to be used successfully. There are, however, general aspects of shark handling which apply to all species.

Sharks do not have bones. Their skeletal structure consists of cartilage. This is easier to remove from the flesh than are the pinbones of bony fish. Consumers appreciate boneless seafood. Flesh color varies from white to dark red, though most sharks have some red meat. In the past, shark meat was considered toxic. Species eaten today were discarded as inedible trash. It appears that toxic incidents from shark had two causes. One was the consumption of shark liver, which contains toxic concentrations of certain vitamins. The other was consumption of meat that had not been handled properly.

Handling shark correctly is absolutely critical to successful marketing and satisfactory use. Sharks carry urea in their blood. Urea breaks down to ammonia. The blood also carries a related substance called trimethylamine oxide, which breaks down rapidly into another unpleasant spoilage indicator, trimethylamine (TMA). It is essential to bleed shark thoroughly, or these substances remain in the flesh, making refrigerated shark smell terrible very quickly and frozen shark smell almost as bad in a surprisingly short time—the enzymatic breakdown process is slowed but not halted by freezing. Fishermen report that blue shark taken in nets that are hauled on board dead often smell bad even before they are on the boat.

The quality and acceptability of shark meat depends initially on the fishing method and then on the effectiveness of bleeding. After that, all the other factors that apply to most other fish come into play, such as the proper use of ice, temperature control and sanitation. Netted shark of most species need to be brought onto the fishing vessel very quickly. Sharks do not have gill flaps, which propel water over the gills of other fish. Instead, they have to keep moving to force water into the gills. If they are unable to swim, they asphyxiate. Consequently, net caught fish are often dead when landed and dead sharks cannot be bled effectively. It is better to use sharks caught on hooks and brought on board the fishing vessel alive. Live sharks should be bled by cutting off the tails. This permits the heart to pump out as much of the blood as possible before the fish dies. Gutting the fish as soon as it is captured does not work nearly so well. It is better to bleed the live fish, which may take as long as 30 minutes and then process and ice it.

Soaking the meat also helps to remove the taint of ammonia and TMA. Lactic acid, citric acids, milk and salted water are possible soaking mediums. Many shark recipes start with an acidic or milk marinade, which is actually designed to remove the ammonia rather than to impart flavor to the meat. Soaking can be quite effective, but it seems that different soaks work better on certain species.

Fresh shark is available as dressed fish or cut into fillets, slabs or steaks. Whiter meat is generally preferred. If it has been properly bled, shark meat has good keeping qualities. If it contains much blood, it will start to smell very quickly.

Opinions vary about which shark are good to eat and which ones are not. Probably, they are all edible, but the handling techniques used are better for species for which there is an established market, such as mako. Fishermen will take care of a high value species but may not know or care about one which is not a regular part of their business. The list in Table 7 offers some guidance but does not claim to be in any way definitive. Neither is it exhaustive: there are over 100 species of shark found in North American waters. These vary in size from the small dogfish (see below) through the basking shark, which reaches over 30 feet in length and 8,000 pounds in weight, to the largest of all fish, the whale shark. The longest specimen of this species was almost 40 feet. The weight of a slightly smaller one was estimated at over 26,000 pounds. It is doubtful if any fishermen would want one of those snapping its jaws while bleeding to death on the deck.

Mako, white tip and black tip sharks are sometimes substituted for swordfish, although mako (which is two species, longfin and shortfin, both found off all American coasts) has earned itself a market in its own right and is sometimes expensive enough that substitution makes little sense. Shark skin is rough, like sandpaper. Swordfish skin is smooth.

Table 7. Suggested edibility of shark.

Generally considered good for the table:

Angel	Sevengill
Blacktip	Sharpnose
Blue	Shortfin mako
Bonnethead	Silky
Common thresher	Sixgill
Lemon	Smooth dogfish
Longfin mako	Soupfin
Porbeagle	Spinner
Salmon	Spiny dogfish
Sandbar	Whitetip

Generally considered poor eating:

Atlantic thresher	Nurse
Bull	Pelagic thresher
Dusky	Tiger
Hammerheads	

The feel of the skin is the most positive way to tell if shark is being shipped instead of swordfish. See also SWORDFISH.

DOGFISH: *Mustelus canis* is the smooth dogfish from the Pacific. *Squalus acanthias* is spiny dogfish, which is found on both east and west coasts. Nomenclature for dogfish is confused, partly because dogfish is a poor marketing name. Other names that have been or are used include dog shark, huss, flake and rock salmon. Gray shark, Cuban cod and other fanciful notions have been suggested. None of these is sanctioned by the FDA, although flake is an approved name in Australia. In 1990, the FDA agreed to allow the use of the name cape shark. This is intended to make it easier to market dogfish, without causing confusion with any other fish.

Dogfish are small sharks. Smooth dogfish grow to about 12 pounds and spiny dogfish to about seven pounds. The single skinless and boneless fillet from the back of the fish weighs between one and five pounds. Fishermen dislike catching dogfish because they are very rough and break nets, as well as being unpleasant to handle. Processors also dislike them because the tough skins wear out processing machinery quickly. The meat is firm, white and sweet and is very good for generic fish and chip meals. However, at present most of the catch is processed and frozen for export markets in Europe.

The bellyflaps are sold to German processors who turn them into a hard smoked product called schillerlocken. Bellyflaps are often worth more than the fillets. The tails are sometimes dried or frozen for use in shark fin soup.

SHEEPSHEAD

FRESHWATER DRUM: *Aplodinotus grunniens*. The only name sanctioned by the FDA is freshwater drum, but in New York it is more often called silver perch and in the south it is called gaspergoo.

It is a coarse fish from freshwater lakes. The sheepshead averages a little under two pounds and is usually sold dressed, as an inexpensive pan fish.

CONVICT FISH: *Archosargus probatocephalus*. Sheepshead is the only approved name for this species, which is also known in some markets as fathead. This is an Atlantic and Gulf fish, growing to as much as 20 pounds, and is popular in Florida and New York. It has good, firm flesh. Most of the sheepshead sold commercially is under 10 pounds and is dressed.

Sheepshead are caught mainly in the Gulf of Mexico in the winter months. The species appears to be abundant. It is inexpensive and might repay the time spent examining it with a view to using it in areas which have not so far seen sheepshead.

CALIFORNIA SHEEPHEAD: *Semicossiphus pulcher.* Note the spelling of sheephead, without the "S" in the middle. This is a wrasse, often around 20 pounds, found off southern California during the winter. Small quantities are sometimes sold commercially.

SHELLFISH SANITATION
Oysters, clams, mussels and scallops are all filter feeding shellfish, which pump water through their digestive systems. Humans eating them run a risk of contracting illness. This risk is small, but is greater than that from eating finfish or other shellfish. There are four reasons for this higher risk.

- We often eat the whole animal, including the guts. This means that we are consuming any bacteria or other microorganisms present in the animal's gut. Some of these microorganisms can be harmful.
- We often eat these shellfish raw, or, if cooked, only very lightly cooked. Thorough cooking destroys most of the microorganisms.
- We seldom freeze oysters or clams before eating them whole and raw, even though freezing can destroy microorganisms almost as effectively as does cooking. We do not use frozen shellfish because of ingrained prejudices, not because the product is in any way inferior.
- Filter feeding molluscs can absorb, without major harm to themselves, toxins harmful to humans. These toxins are naturally produced in the marine environment. Shellfish have to be monitored for the presence of these toxins, which are not destroyed by freezing, by cooking or even by canning.

If filter feeding molluscs are harvested from clean water and handled correctly, there is virtually no risk in consuming raw shellfish. The animals do not absorb harmful microorganisms from clean water, only from polluted water. If they are handled properly, they do not become contaminated during handling and transportation and any small colonies of bacteria that may be in them are not able to multiply to levels where they might cause harm to consumers.

Although *E. coli* is used as the indicator to show whether harmful bacteria may be present, the Vibrio and Cholera groups are among many which are more dangerous. The presence of such bacteria may also indicate the presence of such viruses as those causing hepatitis and polio. As well as microorganisms, the National Shellfish Sanitation Program (NSSP) controls look for marine toxins, such as those causing PSP and domoic acid poisoning.

In 1925, following an outbreak in New York of typhoid caused by people eating shellfish that had been harvested next to a sewage outlet, the NSSP was instituted. This is a cooperative program between the federal FDA and the states. Governed by another body called the Interstate Shellfish Sanitation Conference (ISSC), which sets the rules and procedures and updates them at regular meetings, the NSSP is operated by the states and monitored and policed by the FDA. Simply, stated the NSSP sets standards for growing water and for the way shellfish must be handled to ensure the safety of shellfish consumers. The Program covers all edible species of oysters, clams, mussels and scallops, fresh or frozen, however processed, that are shipped in interstate trade or are imported. (Scallops are excluded if only the shucked adductor muscle is used. All other scallop products are included in the rules.)

The administration of the NSSP depends on two things. One is that all growing waters have to be inspected regularly and must meet designated standards if they are to be approved for shellfish harvesting (the rules apply to shellfish farms as well as to uncultured resources). Shellfish must not be taken from any water that is not approved. This means that areas that have not been surveyed may not be used for harvest. Only tested and approved waters may be harvested. Approval is dependent upon the safe condition of the water at all times. Approved areas that suffer a bloom of red tide algae or an unusual sewage spill, for example, are immediately closed. The second point is that all shellfish processors, shippers and dealers must have a state license, which is awarded only if the licensee meets certain standards for sanitation and handling procedures. These licenses are reported monthly in a free publication called the Interstate Certified Shellfish Shippers List (ICSSL). Every commercial user of fresh and frozen shellfish products should receive this list, which is available from the FDA, Shellfish Sanitation Branch HFF-344, 200 C Street SW, Washington, DC 20204. Never buy shellfish from anyone not currently listed.

Although the NSSP applies technically only to interstate trade, in practice all states enforce similar rules and standards for shellfish sold within their borders.

Obviously, the Program is much more complicated than this brief summary makes it sound. The full detail of the Program is set out in the NSSP *Manual of Operations,* which specifies all the standards and procedures for all aspects of the Program. The Manual is available from the FDA, Shellfish Sanitation Branch HFF-344, 200 C Street SW, Washington, DC 20204. The rules constitute part of the U.S. Code and as such carry the full force of law.

For shellfish users, it is important to understand the principles of the NSSP and to appreciate and follow some simple procedures.

- Never buy shellfish from anyone not listed on the current ICSSL. Everyone who resells shellfish, including local wholesalers who break

down large shipments of shellfish into smaller batches, must have a license from the state.
- All live shellfish must be tagged. The tag shows the origin and the license number of the shipper and the date the shellfish were harvested. If there is a health problem with any shellfish, then other shipments from the same area which might be affected can be tracked down quickly from the information on the tags. If a bag or other container of shellfish arrives without a tag, refuse it. Even if the tag is on the floor of the truck and you are "sure" that it came from your bag of shellfish, you should not accept the product. This is not just good sense, it is required by the law.
- The final vendor of the shellfish—usually the restaurant or the retail store—must keep the tag for at least 90 days. In case of later liability claims, it is sensible to keep it for many years. The rules require that you keep the tag, or at least the detailed information on it, for at least three months and make that information available to health officials when required.
- Shucked and processed fresh and frozen shellfish must carry the license number of the shipper on the container. As with the tags, the final receiver must keep this information on file (it is simple to associate and keep the records with the invoices) for a minimum of 90 days. As with tags, never accept product that does not bear the required information.

While the NSSP has been successful in protecting both consumers and the industry that depends on them, there are problems facing the system. No rules and procedures can be effective unless people understand and follow them. The following paragraphs outline a few of those problems, not in any order of difficulty or importance.

- The classification of growing waters, which is absolutely vital to the success of this or any similar program, is expensive and difficult. There are huge areas of potentially productive shellfish waters around U.S. coasts which are not surveyed and therefore are closed. The cost and manpower to survey all possible growing waters is simply not available. Consequently, shellfish harvests are limited by the ability of the states to maintain their assessment of growing areas.
- Illegal harvesting from waters not approved for shellfish is a continuing problem. Enforcement is difficult. The first buyer of shellfish must ascertain that the product was legally fished; however, if the fisherman says it was, there is no way the dealer can argue. Closed areas often produce a lot of shellfish, simply because no one is fishing them. The temptation to harvest such areas is, unfortunately, too much for some people. Illegally harvested shellfish can be dangerous

and puts the whole industry at risk. Funding for enforcement is limited and penalties are generally light.
- Imported product is supposed to be subject to similar controls. Before shellfish can be shipped to the United States from overseas, the FDA must reach an agreement with the foreign government, which undertakes to control harvest areas and handling sanitation according to the same standards as those in the NSSP *Manual*. Only shellfish from approved areas may be shipped to the United States. At the present time, there are agreements with only nine countries, including Canada. It appears that shellfish from many other countries does enter the United States. Some of this may be misdescribed. Customs officers may not realize that cockles count as clams. The sheer number and complexity of the country's imports contributes to the difficulty of enforcement. For whatever reasons, importation of shellfish from unlicensed sources overseas is a potentially serious problem for the industry and for American public health.

In summary, therefore, the system is in place to ensure that the shellfish reaching the consumer is as safe and wholesome as it reasonably can be. The system works only as well as those affected make it work. While almost all of the shellfish industry understands and appreciates the importance of the NSSP, ignorant or maverick fishermen, dealers or importers could destroy the system and even the industry by causing a few well publicized instances of serious illness from illegal shellfish. For a full discussion of the NSSP and information on the pathogens borne by shellfish, see this author's *Shellfish—A Guide to Oysters, Mussels, Scallops, Clams and Similar Products for the Commercial User*, published by Van Nostrand Reinhold/Osprey Books.

SHRIMP Although shrimp is, in dollar terms, by far the most important seafood item and is the staple for many dealers and distributors, comparatively little of it is sold fresh. Frozen shrimp is the normal and accepted product throughout the distribution chain. Fresh shrimp is available in coastal production areas. Specialized items such as whole (or even live) freshwater shrimp are sold in very small quantities.

NORTHERN SHRIMP: *Pandalus spp*. These are cold water species. Vernacular names are numerous, including many of the geographical origins. The major commercial species is *Pandalus borealis*. This is a circumpolar shrimp, found from polar regions south to Britain in the eastern Atlantic, Martha's Vineyard in the western Atlantic, Oregon in the eastern Pacific and Honshu and South Korea in the western Pacific. The largest numbers are in Scandinavia, western Greenland, the Gulf of Maine, Oregon and Washington. The northern shrimp grows to about 6 inches.

SHRIMP

Northern shrimp is small, but has excellent flavor and texture. It is difficult to peel when raw, so it is generally sold cooked and peeled. Although most of it is sold in frozen form (or thawed), some is sold fresh. The cooked shrimp tails are pale pink, moist and well flavored. The meat is softer than that of the more familiar tropical shrimp. Whole cooked shrimp are also available, for the small minority of consumers who are willing to peel the shrimp themselves. Yield is very small, but the meat seems to retain flavor and succulence better inside the shell. Whole cooked northern shrimp is highly susceptible to melanosis. Blackening appears on the heads in a very short time. Unless severe, this is a natural feature of the shrimp and should not be a reason for rejection.

Since northern shrimp is invariably cooked, bacteria growth is a particular risk. Fresh cooked and peeled shrimp must be handled under the most hygienic conditions and should be used very quickly. It is probably unwise to keep it for more than 48 hours after cooking, even under ideal conditions where it has been chilled quickly and kept cold and clean. If it is displayed in a retail case, make sure that it cannot be contaminated with moisture or bacteria from raw fish and shellfish nearby. Cooked shrimp (and other cooked product) should be handled with utensils reserved for them, to avoid any risk of transferring bacteria from raw or live fish. The taste of fresh cooked northern shrimp is certainly worth the care and effort it takes to supply a good, wholesome product.

There are a number of related pandalid shrimp species that are occasionally offered fresh in the United States, especially in the Northwest, where most of them are fished. Of these, spot shrimp (*Pandalus platyceros*)—spot prawns in Canada—are probably the most common. This species grows to 10 inches and is found along most coasts of the northern Pacific. In the United States, it is usually sold whole and cooked. Because of its large size, it fetches a substantial premium over northern shrimp.

TROPICAL SHRIMP: Shrimp from warmer, but not necessarily tropical, waters supplies the bulk of the U.S. shrimp market, which is the second largest in the world, after Japan. Several species are now successfully farmed. Shrimp aquaculture is a major industry in countries as far apart as Ecuador and China. The United States does not have many areas suitable for shrimp farming, so the domestic industry, which supplies what little product is offered fresh, is mainly a capture operation. It is centered in the Gulf of Mexico. White shrimp (*Penaeus setiferus*), pink shrimp (*Penaeus duorarum*) and brown shrimp (*Penaeus aztecus*) are produced. Any of these may be offered fresh from time to time, mainly in areas close to the fishing ports.

Mostly, smaller sizes are landed and sold with the heads on. Heads-on shrimp look large, since almost two-thirds of the length of the shrimp is head and consumers are readily convinced that they are buying shrimp much larger than they are actually getting: frozen shrimp counts are based on the number of tails to a pound, not the number of whole shrimp.

Frozen shrimp is not only widely acceptable, but it is also superior. Quality control, branding and all the other factors which make for a reliable product can be applied to frozen shrimp. Fresh shrimp tends to be an opportunist's business aimed at quick profits and no permanency, except for local exceptions in producing areas such as Louisiana and Florida.

If you must use fresh shrimp, examine every container thoroughly to make sure the product is really fresh. The shells should be firm and shiny, and should feel full of meat. Softness often indicates staleness. Black spot (melanosis), which is the appearance of black spots and patches, especially on the heads, is a sign of stale shrimp (see MELANOSIS for a full explanation). Check the counts by weighing one pound of product and counting how many shrimp you have.

As the costs of farming shrimp decline and the technology spreads even further, it is possible that there will be a trend towards airfreight shipments of fresh shrimp direct from farms overseas. Because fresh shrimp has a limited shelf life, such operations will require considerable care and control.

LIVE SHRIMP: There is a very small market for live shrimp, which can be transported by air in specially designed containers. This is a specialized, high risk business. Shrimp can, with skill and luck, be kept alive in tanks. More often, they are sold freshly dead. As aquaculture techniques develop, pond shrimp (grown in fresh water, not sea water) may become more widely available. The giant freshwater shrimp (*Macrobrachium rosenbergii*) is a very large species, reaching over 12½ inches in length. Alive or dead, the whole shrimp look impressive. The species is available from farms in Hawaii.

SHRIMP COUNTS: Fresh and frozen shrimp tails are size graded and sold by count per pound. Use of verbal size descriptions instead of count used to be sanctioned but not recommended by the National Marine Fisheries Service. That limited approval has now been withdrawn and it is probable that use of size descriptors for shrimp constitutes misbranding. Words such as jumbo, colossal and large should not be used. At one time, these words were defined in a U.S.D.C. Grade standard, but the definitions were much abused. Size descriptions for canned shrimp are still permissible under FDA regulations.

SKATE *Raja spp.*
Skates are also called rays, although rays are actually different fish. Skates are abundant in all waters of the United States, but the small quantities landed are mostly exported. Small markets exist in New York and other major eastern cities. Skate is prized in many European countries.

The edible part of the skate is the wing on either side. Two fillets can be removed from this, or it can be used whole if the fish is small enough. The underside usually has white skin and the upper side brown skin. The upper fillet is thicker than the lower one. Remove the skin before cooking the fish. Wings sometimes come with the knuckle joint, which attaches the wing to the body. This is also normally removed before the wing is cooked.

There are many skate species and they vary enormously in size. Small wings may weigh as little as eight ounces. Large ones may be over six pounds. Skate meat is sweet and firm. Fresh skate can be used for various special dishes from European cuisines and also for fish and chips. It does not keep too well, developing an ammonia taint unless carefully and rapidly handled. Like shark, this fish contains urea in the blood and does not excrete. The urea turns to ammonia quickly. Some recipes suggest marinating skate in milk or citrus juices before cooking. This will help to remove any ammonia taint. Bleeding the fish as soon as it is caught is the best solution but is seldom if ever practicable because of the conditions when the fish is caught. Also, it is normally a by-catch, considered a nuisance rather than an asset.

The common belief that imitation scallops can be punched out of the skate wing with cookie cutters pays little regard to the high costs that such a process would incur. However, it does indicate that the flesh quality is good, if it might possibly be mistaken for a scallop. These fish offer cheap raw material to users willing to take the necessary trouble and care in buying and handling them.

SLIPPER LIMPET   *Crepidula fornicata.*

These small snail-like molluscs attach themselves in large numbers to the shells of oysters and mussels. They breed rapidly and use the same food as the oysters and mussels, thus competing with them. They are therefore considered a nuisance by shellfish farmers. They can be boiled or eaten raw, though there is not much meat. The market is limited to curiosity seekers. However, if a market could be developed, the species would no doubt decline and the resource become scarce, thus improving the situation of oyster growers.

SMELT   *Osmeridae.*

Also called whitebait, argentina or argentine, eulachon, hoolihan, candlefish, capelin, silverside, sandeel, lance. Most of these alternative names are incorrect or unacceptable, but it seems likely that they will continue in use. Smelts are small, silvery fish from northern waters of both the Pacific and Atlantic oceans. Some are anadromous and others live entirely in freshwater. The argentina is a deep water sea smelt with rather soft flesh which is being examined for industrial exploitation by European fishing companies. Although all species of this fish are different, in use most of them are very similar.

The major smelt runs are in the northeast and eastern Canada, the

Great Lakes and the Pacific northwest. Spawning fish run into rivers or close to shore and can be caught easily in large numbers.

Smelts are eaten whole, if very small, as whitebait. More often they are headed and gutted, with the guts removed through the neck opening, washed and then used. Fried or broiled, the flesh is generally bland and tender. Bones may be eaten in smaller fish, though the backbone is easily removed in one piece once the fish is cooked. Smelts about five to seven inches long are preferred. At this size, they run about 12 to the pound. Smelts seldom exceed 12 inches long.

Lake smelts, many of which are imported from Canada, are considered to have a finer flavor than sea smelts. They are a little less oily, as a rule. (The candlefish of the northwest got its name from being used as a candle, there is so much oil in the meat.)

Smelts lose flavor rapidly as they age and are best eaten very soon after they are caught. For that reason, frozen smelts may often be a better choice than fresh.

Capelin is a smelt and is sometimes used as one. See the separate heading on CAPELIN.

SMOKED SEAFOODS

Heavily smoked and salted foods were important in past centuries since they remained edible for many months—a vital feature before canning, refrigeration and other modern forms of preservation. Many of these smoked products were probably not very palatable and would certainly not appeal to modern American or European tastes. Compare modern smoked products with a true red herring which was smoked for about three weeks, giving it plenty of time to be fully cured even if the conditions inside the kiln varied greatly from day to day. (Red herrings date from at least the 16th century. The fish were smoked whole, complete with guts and head. The guts were eaten with the rest of the fish.)

In fact, most of the so-called traditional smoked products still made were developed during the 19th century and were much lighter cures than the original smoked foods. Kippers, for example, were first produced in northeast England (at Seahouses in Northumberland) in 1843. The point of these new products was to supply foods with an attractive taste and texture. This development has continued and modern smoked goods, for the most part, are produced for their attractive flavor. Preservation is no longer a consideration and the light cure of most modern products gives them no more shelf life than unsmoked product. The rule to follow with smoked seafoods is that they should be handled and treated just like fresh seafoods.

Modern smoking kilns offer considerable advantages over traditional methods. Old kilns were basically a chimney in which the product was suspended in smoke. Consistent control of the density of smoke and of the temperature was not easy. Control of humidity, which is vital, was very difficult. Control of these factors is most important for modern products which are smoked for only a few hours and must be consistent in quality and flavor every time. Modern kilns are fully automatic and carefully monitored and controlled. Nevertheless, there remains considerable skill in smoking because of the great variability of the raw material. Judging the quality of smoked fish is easier if you know something of the processes and the principles involved.

Good product can only be made from good fish. Stale fish does not smoke well. The staleness reappears within a day or two of smoking. There is a widespread belief that smoking is a useful way to salvage or disguise stale fish. This is not so.

Fish and shellfish also have to be in suitable condition for smoking. Herrings must have a high oil content—for example, herring that has recently spawned will give at best a dry and rather hard product. Bruises in fish flesh show up clearly after smoking. This can be a particular problem with salmon, causing smokers to prefer line caught (trolled) fish over netted fish since the hooked fish is less likely to be bruised. Similarly, inadequate bleeding can give a discolored product. In general, good smoked product must be made from the highest quality raw materials. The definition of high quality relates both to the intrinsic biological condition of the fish and to the way in which it was handled and processed.

The first step in the smoking process is to salt the fish. Salt preserves flesh by replacing some of the moisture in the tissue, making it less hospitable to spoilage bacteria. Salt also, of course, brings out the flavor. Traditional salting processes used dry salt, but most modern products are brined in saturated salt solution. Sometimes, dyes are added to the brine. Fish that tastes bitter may have been brined in the wrong type of salt: calcium and magnesium sulphates which are frequently found in salt have this effect. It is important that the brine is kept clean and at full strength. Blood, scales and soluble proteins from the fish mix with the brine as it is used, so the solution must be regularly renewed.

It is possible to put the fish through a second brine containing spices if special flavorings are desired. Some processors add nitrites, mainly in the form of sodium nitrite, to their brines to add a desirable color to the fish and help keep it moist. Federal regulations limit nitrites in the finished product to 200 parts per million. Too much nitrite is extremely poisonous.

After brining, the fish should hang in a gentle flow of cool air. This allows excess brine to drip from the flesh and gives time for a glossy surface to develop on the cut surfaces of the flesh. This gloss is both attractive to the consumer and protects the product a little. The gloss is called the pellicle.

Smoking is actually a combination of two separate processes: drying and smoking. Removal of some of the moisture from the fish is necessary to give the product the required texture. During smoking, chemicals are deposited on the flesh. These chemicals contribute to the flavor of the fish and help (a little) to preserve it, since some of them kill surface bacteria. Preservation depends partly on drying as well; in general, the more moisture remaining in the flesh, the shorter the shelf life is likely to be. Modern products are mostly rather moist. Consumers prefer them that way and processors lose less weight. The trade off is loss of the preservative qualities of smoking.

Hot smoking and cold smoking techniques produce different products. Cold smoked products are processed so that the temperature of the fish remains below 85°F. Many cold smoked products need to be cooked before being eaten. Hot smoked products are cooked in the smoke, usually by increasing the temperature to 140°F to 150°F to complete the operation after an initial period of cold smoking.

Smoking and drying both occur in the smoke, but are separately controlled. The amount and density of smoke, together with the type of wood dust used, determine the flavor of the product. The relative humidity of the air in the kiln determines the amount of drying.

The loss in weight during smoking can be considerable. In general, it is between 15 percent and 30 percent. In considering operation of a smoking kiln, the expected weight loss may be the most expensive cost of production.

There are several liquid smokes or smoke dips available, which can be used for some items instead of true smoking. The texture of the finished product is the same as for fresh product, but items cooked with a liquid smoke can be very good. Generally, using liquid smoke produces a barbecue rather than a smoke effect. Liquid smoke can also be injected into the flesh, producing an imitation smoked product in a remarkably short time.

Handling smoked products is similar to handling other fresh seafoods. It is necessary to keep them cool. Ice should not be allowed to come into contact with the flesh, however. For good hygiene, smoked fish must never be displayed or stored where drip from raw fish or shellfish might contaminate it; this will cause rapid bacteria growth and can lead to food poisoning. Because the smoking processes kill many of the bacteria which cause odors, it is difficult to detect when smoked fish has gone bad. Consequently, proper refrigeration is particularly important for smoked product. Some product is canned or packed in retortable pouches, which need no refrigeration. Both these forms of packaging, however, involve cooking the product in the container, so taste and texture are inevitably different from those of the fresh smoked product.

Freezing smoked fish is acceptable provided the product was made with the intention of freezing it. Saltiness is enhanced by freezing, so it is necessary to brine less if frozen product is required. Frozen smoked products are generally regarded as inferior to fresh smoked product,

SMOKED SEAFOODS

although, as always, a product that was properly frozen and handled is preferable to an inferior "fresh" product.

Vacuum packing smoked fish is now common, despite some potential risks of growth of anaerobic bacteria (such as the one causing botulism) on the low-acid surface of smoked seafoods. Vacuum packing (and the related technique of retortable pouches) has been successful and safe because processors take care to ensure sanitary conditions and to use films which allow oxygen to pass in small quantities. Vacuum packs must not be abused in distribution. It is very important to keep the product cold throughout its life. If vacuum packed smoked fish should warm up, the pack should be opened so that air can get to the surface of the fish. This will remove any risk of multiplication of any anaerobic bacteria that may be present. Similarly, when frozen product in vacuum packs is to be thawed, the packs must be punctured. Vacuum packs increase the shelf life of smoked seafoods but do not avoid the necessities of handling the product cleanly and quickly and keeping it cold.

Presentation of smoked product is important, especially with the most expensive items like smoked salmon. Thinly sliced salmon can be laid out on a fish shaped board for a sumptuous appearance which uses only a few ounces of product, making it both affordable and highly desirable. Luxurious packaging is most important in the presentation and sale of such items. French smokers have for a long time had some excellent packs which U.S. distributors are beginning to imitate.

The following are notes on some individual smoked products, listed in alphabetical order. There are many other seafoods which can be smoked and sometimes are smoked: smoked cod roe is popular in parts of Europe; shrimp, shark, halibut, butterfish, bluefish and many other fish make products varying from good to excellent. At some time or another, almost every seafood seems to have been smoked and a surprising number of them are attractive products. Popular taste seems to be increasingly approving of smoked seafoods.

BLOATERS: These are popular in Britain and eaten in Canada, but seldom seen in the United States. Bloaters are herrings which are dry salted whole and then cold smoked for a short time. The product retains the silvery color of the raw herring and has a distinctive taste, probably from the fermenting of the contents of the gut. Smoked chub may sometimes be called a bloater. There is little similarity between the products.

BUCKLING: Hot smoked herring, sometimes whole but more usually gutted. Buckling originated in Germany. The fish have an attractive golden brown color and are fully cooked. The flavor is rather strong.

COD AND HADDOCK FILLETS: These are largely interchangeable in practice, though since most of the product is described as smoked

SMOKED SEAFOODS

haddock, it can be assumed that haddock is more desirable. The two can be distinguished by the haddock's thumb mark on the skin near the head, which the cod lacks, and the black lateral line. Fish without the skin will need scientific analysis to be distinguished.

Fillets are colored and brined, and then cold smoked. It is possible to produce an acceptable product by using a liquid smoke dip or injection system, which has the advantage that no weight is lost in the production process.

EEL: Smoked eels are not much appreciated in the United States, but are a great delicacy in Germany, Holland and other parts of Europe. Eels are exported alive from the United States for smoking in Europe. Few are processed and sold here. Eels are hot smoked and ready to eat. Small eels of about one pound as well as large eels weighing up to 10 pounds are used for smoking.

FINNAN HADDOCK: Also called finnan haddie and smoked haddock. Properly, a finnan is a small haddock that has been split along the backbone, leaving the backbone attached to the flesh of one side. The fish is lightly brined and cold smoked. It should be an attractive golden color. The term finnan haddie is widely used to describe any smoked haddock or cod, including fillets which are preferred by American consumers. See above for comments on smoked cod and haddock fillets.

HERRING: This is the raw material for many smoked products, including bloaters, buckling, kippers and blind robins. As mentioned, it is important to use herring with a high oil content.

KIPPER: Herrings are split along the back, brined and cold smoked. Color develops slowly, after smoking is completed: dyed kippers may turn very dark, even black. (Canned kippers may turn black simply from an excess of smoke, without any dye being used.) Kipper fillets are more popular than regular kippers in the United States. It is important that the herring for kippering should have an oil content of at least 10 percent, and preferably far more. Low oil content herring makes a dry and unsatisfactory product.

MACKEREL: Smoked mackerel similarly requires an oily fish. The fish are generally nobbed (heads cut off and the guts drawn through the neck opening) or headed and gutted, brined and hot smoked. They are ready to eat and make an excellent alternative to more expensive smoked trout.

OYSTERS: Oyster meats can be lightly brined, dipped in oil and hot smoked. Generally, the product is then preserved in cans or jars. Similar use can be made of mussel and clam meats.

SMOKED SEAFOODS

POLLOCK: An imitation smoked salmon product was produced in Germany during the First World War. Known as seelachs (sea salmon), it is made by heavily salting large fillets of pollock, slicing the fish and soaking the slices in a mixture of brine, acetic acid and dye. The slices are then processed in a very dense smoke, packed in vegetable oil and either used fresh, with a shelf life under refrigeration of about one month, or canned.

SABLEFISH: Filleted, hot smoked sablefish is an excellent product, well known on the west coast and in major cities, but by no means universally available. Like fresh sablefish, it deserves to be better known and appreciated.

SALMON: Smoked salmon is probably the best known and most popular smoked fish product. The original product from northern Europe is made from a fillet of Atlantic salmon heavily brined and then cold smoked for lengthy periods (the exact time depends upon the size and fat content of the fish). Scottish, English, Irish, Norwegian, Icelandic, French and Danish smoked salmon are all made in similar fashion. Each smoker has his own special techniques and works in a particular way, often giving a certain individuality to his product. Pacific salmon has been smoked in Europe in exactly the same way for 150 years; before refrigeration, the salmon were shipped pickled in barrels of brine.

Although cold smoked, the salmon is ready to eat without further processing. The hardened pellicle must be discarded and the pinbones removed one by one with long nosed pliers. The side is then sliced very thinly and eaten. Removing the bones is important because although they are soft and can be cut and eaten, the presence of the bones usually causes the slices to tear when being cut, which spoils the appearance of the product and is wasteful. Sliced smoked salmon products are usually frozen and then mechanically sliced. The slices may remain frozen or may be thawed and sold as fresh. If the fish is sliced when frozen, it is not necessary to remove the pinbones.

Nova salmon is a term used on the east coast to describe a product very similar to the European smoked salmon defined above. The name may have originated from Nova Scotia, but most production now comes from New York City. Nova salmon tends to be slightly less salty than European and varies in quality from very fine to rather poor.

Lox, strictly speaking, is cured, not smoked, salmon. Although originally salted sides, the cure is now lighter and includes sugar, making a mild product. Lox is soaked in fresh water to remove most of the salt before it is eaten. It is sliced and used much the same as smoked salmon. However, most of what is described and sold as lox is actually lightly smoked salmon, since the popular meaning of the term seems to have embraced the smoked product.

Cold smoked salmon is sometimes packaged in retort pouches. the retorting cooks the fish, changing the texture entirely. This can be an excellent product, but it cannot be sliced and used like regular un-retorted cold smoked salmon. Instead, it is cooked salmon with a pronounced smoked salmon flavor. This packaging is mainly used in the northwest and has particular appeal for product sold as gifts and through the mail.

Kippered salmon is fillets or pieces lightly brined and then hot smoked. Freshly prepared, hot smoked salmon is excellent if it has not been overcooked, which dries it. Hot smoked salmon is often packed in retort pouches. It is a little drier than cold smoked salmon similarly packed.

Indian cure, hard cure salmon or salmon jerky is a west coast product originally used to preserve the fish for long periods. It is heavily brined and then cold smoked for a long time, around two weeks, giving a product like fishy pemmican. This product seems to be increasing its following.

TROUT: Smoked trout is an excellent gourmet item. It is hot smoked, ready to eat and makes an impressive appetizer. Trout are smoked with the heads on, since they are hung in the kiln and spearing the edible parts of the fish would tear the flesh and spoil the product. Most consumers prefer the heads removed before serving. Trout should not be dried too much while being smoked as they have a comparatively low oil content and overdrying results in a hard textured product.

WHITEFISH, CHUB, CISCO: These freshwater fish are all excellent hot smoked. Main markets are in the Midwest and in the Northeast, especially New York. Smoked chub are sometimes called bloaters.

SMOLT
A young salmon when it first leaves its home river and enters the sea. Reliable production of smolts by hatcheries is the cornerstone of the salmon farming industry. Without smolts, there can be no adult fish to send to market. Tracking the numbers of smolts put into sea pens can be a useful way of assessing future production trends from salmon farms.

SNAPPER
The second most abused word in seafood nomenclature. This section is about snappers (Lutjanidae), which are expensive, good quality fish, generally with red skin. Red snapper is the benchmark for the group. Fish called Pacific red snapper are covered in ROCKFISH. These species are quite unlike red snapper in taste and texture, although they do have reddish skins. The red snapper name is used to misdescribe all sorts of fish, from ocean perch and grouper to tilapia. The snapper name without the adjective may also be applied, without intent to mislead, to certain species of turtle as well as to small bluefish, cod and haddock. In New Zealand, a sea bream is called

snapper (but take care not to try to import it under that name). Any skinless fish offered as (red) snapper is undoubtedly something else: the skin is removed to prevent positive identification by the receiver.

There are plenty of imported snappers. Brazil, Thailand and Taiwan have well established markets for quite different types of frozen product. Snappers are quite common in the Arabian Gulf and other regions around the Indian Ocean. Some of these are flown into the United States for sale fresh, usually gilled and gutted. Identifying these fish requires expert assistance and plenty of time. There are plenty of snappers but only one red snapper. The FDA issued a Compliance Policy Guide in 1980, setting out the background to the problem. FDA policy was clearly stated:

> The labelling or sale of any fish other than *Lutjanus campechanus* as "red snapper" constitutes a misbranding in violation of the Federal Food, Drug and Cosmetic Act.

Apart from red snapper, there are plenty of other legitimate snappers. The *Fish List* has 37 species which may be called snapper. Every one of these species requires a qualifier in the name, such as yellowtail snapper, queen snapper and mutton snapper. In addition to red snapper, one species, *Lutjanus purpureus,* is called the Caribbean red snapper. See the Appendix for the edited *List*.

Many of the snappers are very similar in appearance, taste and texture. It is not clear why red snapper enjoys the especial esteem of consumers, but the fact is that it does. Other snappers may be as good to eat, but if the customer is willing to pay extra for a particular species, he should at least be given what he is being charged for.

All the commercial snappers come from the Gulf and the Atlantic, with most domestic fishing done in Florida. The following is a brief summary of features of some of the more important species.

RED SNAPPER: *Lutjanus campechanus.* An older scientific name was *Lutjanus blackfordii.* This fish is found in the Gulf of Mexico, the Caribbean and adjacent Atlantic waters. It is a red colored fish growing to as much as 35 pounds, although most commercial fish are much smaller, around eight to 10 pounds. The premium commercial size is between two and four pounds, giving a natural fillet of about 10 ounces. The species is distinguished from other snappers by the color pattern of rose red skin on top, becoming lighter towards the underside. It also has a longer pectoral fin than other snappers. The black spots typical of many snappers are present on smaller fish, often faint or missing on older and larger fish. The fins are red, the dorsal fin edged with orange.

At one time, fishermen used live wells and brought them to shore alive. Modern fishermen use ice, which is cheaper and more reliable.

Florida requires that red snapper be sold with the skin on. In practice, most of the fish is sold drawn with the gills left in. If it is filleted, natural fillets are preferred, these being the whole fillet from one side of the fish. Fillets that are cut into two or more portion sized pieces are called cuts and are less desirable because they are thicker than natural fillets and thus considered harder to cook perfectly. Snapper has firm flesh from and excellent shelf life.

Red snapper resources are under heavy pressure from both commercial and recreational fishing. Quotas are being sharply reduced and the fish is likely to be in very short supply for a number of years. Imported red snapper from Mexico and Central American countries is available, but is not sufficient to meet market demand. Generally, the quality of the imported product is good. Gills are normally removed from the imported fish to delay spoilage during shipping.

The following brief descriptions (in alphabetical order) of various Atlantic and Gulf snappers may help in identification. As mentioned above, if the skin is missing you can be sure that the fish is inferior. You need the skin at least to be able to name the fish, unless you wish to take a sample of the flesh and have it identified by iso-electric focusing. One possible indicator is the way the skin behaves when the fish is cooked: true red snapper skin remains flat, but the skin on some other snappers tends to curl. If the skin is scored when you buy the fish, this has been done to prevent the skin curling and is a sure indication that the fish is not true red snapper.

BLACK SNAPPER: *Apsilus dentatus* is rare in the United States and is distinguishable by its dark purple skin.

BLACKFIN SNAPPER: *Lutjanus buccanella* is a larger fish, reaching 30 pounds and 24 inches. It is inferior eating but often marketed as red snapper, being bright red, with a silver belly and no black spot. Black pectoral fins are the distinguishing feature. The skin on this snapper is particularly prone to curling and toughening when the fish is cooked. The name hambone is not approved by the FDA.

CUBERA SNAPPER: *Lutjanus cyanopterus* is the largest snapper, reaching 100 pounds. It is greenish gray, sometimes with a reddish tinge and has no black spot. The larger fish are not particularly good quality eating.

DOG SNAPPER: *Lutjanus jocu* is commonly around five pounds but reaches as much as 20 pounds. It is a dark greenish brown color, lighter and redder below. There is no black spot on the sides and the fins are orange colored.

SNAPPER

GRAY SNAPPER: *Lutjanus griseus* is the commonest Atlantic snapper and is good eating. It is often called mangrove snapper, although the FDA apparently disapproves of the name. It reaches about seven pounds and averages two to five pounds. It has similar coloring to the cubera, being gray on top and fading lower down, with a reddish tinge and no black spot. It is found in brackish water as well as in the sea—the mangrove name refers to its habit of living in mangrove swamps along the coast. Gray snapper from the Florida Keys tends to be a more bronze color, generally lighter than the mainland fish.

LANE SNAPPER: *Lutjanus synagris* is small—seldom bigger than 12 inches—and good eating. It is used mainly as a pan fish. Rosy red in color, lighter below, it has yellow stripes along the body and a black spot on each side.

MAHOGANY SNAPPER: *Lutjanus mahagoni* is not often seen in the United States, but is frequently eaten in the Caribbean countries. It grows up to about 20 inches, has reddish brown skin, silver color around the belly and a black spot on each side.

MUTTON SNAPPER: *Lutjanus analis* is quite abundant and reaches about 25 pounds, with most commercially caught fish being five to ten pounds. It is excellent eating, with very white flesh. Small fish look rather like lane snapper; large fish look very much like red snapper. It has a dark, almost greenish back. The skin on the sides is red, a little paler below. There is a black spot on each side and the fins are red. It can be distinguished from red snapper by various fin characteristics, which are not easily examined by anyone not trained as a marine biologist.

SILK SNAPPER: *Lutjanus vivanus* is fairly common and often sold as red snapper. It is excellent eating. Growing usually to about 10 pounds, it is rose red, with yellowish skin below. Younger fish have a black spot each side, older and larger fish frequently do not. Fins are pink, sometimes edged with yellow.

VERMILION SNAPPER: *Rhomboplites aurorubens* is sometimes called the beeliner, again a term that the FDA does not like. It is very similar to the silk snapper and may also be substituted for red snapper on occasion.

YELLOWTAIL SNAPPER: *Ocyurus chrysurus* is normally around two pounds, although it grows to five or six pounds. It is fairly common and very good eating. It is a dark greenish brown color with yellow spots; the underside has a purple tinge. Fins are yellow. There is no black spot on the skin. There is a distinct yellow line along the side of the fish.

It requires a great deal of experience to distinguish between some of these fish, especially as the colors, markings and the important black spot all fade slowly once the fish is caught. The older the fish, the harder it is to distinguish the different features.

SOLE See FLOUNDER AND SOLE.

SPENT Fish which has spawned. Often, such fish will be in poor condition for eating, with watery, soft flesh. The fat content of oily fish is at its lowest immediately after spawning.

SPIKING A Japanese technique, called iki-shime, for slaughtering high value fish. The fish is paralyzed with a spike in the brain. It ceases to struggle and so does not use up glycogen. It can then be bled. The process yields fish of exceptionally high quality.

SPOT *Leiostomus xanthurus*. A small Atlantic croaker, generally between eight and 24 ounces, caught from Cape Cod to Mexico, with the bulk of landings currently coming from the southern Atlantic states. Spot is easily distinguished from croaker by the large black spot just behind the gill opening. The fish is abundant off Virginia in late summer and is caught off the Carolinas and Florida in the fall. It is estimated that huge quantities are available, but the market is limited.

Spot is used as a panfish, usually sold whole or dressed. The flesh is rather coarse and has a strong flavor. There is a large section of dark meat along the lateral line which is particularly strongly flavored and the oil content of this dark meat reduces the shelf life of the fish, whether fresh or frozen.

It is possible that spot can be used for surimi. Its use directly as a food fish, however, is limited by the rather poor taste and texture of the meat.

SQUID Also called calamari. Squid are molluscs, although the shells have evolved into a small bone or pen inside the mantle (the body). Like octopus and cuttlefish, they are cephalopods. They have eight arms and two tentacles. Squid have fleshy wings along the narrow end of the tube or cone shaped body. Different species range in size from an inch or so to the Chilean giant squid, which is six feet long and weighs 100 pounds. Authorities vary about the size of deep sea monster squid, quoting lengths of up to 60 feet. Whatever the truth, these creatures may be enormous. Commercial use is limited to squid of up to a couple of feet in length. They are found in all seas and although a great deal of squid is eaten (not much of it in the United States), there appears to be resources available for exploitation in various oceanic regions. Squid is highly nutritious, easily digested and could be readily

SQUID

available. Most of the domestic catch is exported. A great deal of the squid sold in the United States is frozen. Fresh squid is available seasonally from California and along the Atlantic coast.

Squid has excellent shelf life if handled with reasonable care. Leaving the viscera in obviously reduces the shelf life. Squid is little affected by freezing—the texture is unchanged even if the flesh is frozen and thawed a number of times—so there is little reason to hold squid fresh for long periods when frozen product will be easier to keep. Fresh squid should have a sweet smell. It becomes rather pungent with age. Skin spots and color are not reliable guides to quality. The pen becomes harder to remove as the squid ages, but this is a test requiring some experience with the species.

The tentacles, wings and mantle are all edible. There is a thin, soft skin covering the body and wings which is usually removed before cooking. The skin changes color dramatically after the squid is caught. This is a natural process and is not a symptom of decomposition. Squid may be "bleached" by soaking it in iced water. This reduces the color changes and whitens the skin. Removing the skin is not particularly easy. Rubbing it off while holding the squid under running water seems to be the easiest method. Processors use machinery to remove the skin.

Clean squid by cutting off the head and tentacles (the tentacles should be retained separately) and removing the guts and pen through the neck opening. This leaves a tube, closed at one end, with the wings attached. It is not easy to remove all the viscera this way. Some processors prefer to slit the tube along its length and then all the guts can be scraped off easily. However, many recipes require the tube to be left whole or be cut in rings, so this easier method of cleaning is precluded. Squid is often shipped whole, leaving the user to clean it. Cleaned squid is more expensive.

Squid must be cooked very gently or it toughens. Use low heat for cooking, as well as short cooking times. Some species toughen more than others, but this rule applies to all squid. Alternatively, squid may be cooked for a long time—say about one hour—when the flesh will have broken down considerably and again be tender. Most of the flavor will be removed this way. Squid is not strongly flavored and usually has less fish flavor than clam strips, which are a popular consumer item. Breaded squid rings are widely available frozen and are a popular item. These are easy to use, whereas producing your own is a lot of work.

The three main types of squid available fresh on the U.S. market are described below. There are many other species available, usually frozen, from many parts of the world.

NORTH ATLANTIC LOLIGO: *Loligo peallei.* Also called long finned squid and trap squid. The wings are about half the length of the body tube. Regarded in the northeast as the superior squid, most of this is caught between Georges Bank and Cape Hatteras, though the species ranges as far south as the Gulf of Mexico. It is caught mainly inshore in the spring and summer months, although large quantities are also caught further offshore during the winter. Traditionally, large quantities were caught in traps in Narragansett Bay and southern Massachusetts. Few of these traps still operate, but squid from them is the best available as the fish is alive when taken from the water and can be handled with the greatest care.

Long finned squid has very white flesh and is tender unless overcooked. Tube lengths range up to about 10 inches for the most part, though very large and very small squid is also found.

NORTH ATLANTIC ILLEX: *Illex illecebrosus*. Also called short finned squid and summer squid. The wings are about a third the length of the body tube. Caught off the northeast and middle Atlantic coasts, illex is generally larger and coarser than loligo and usually sells for about one-third or less of the loligo price.

Illex is rather tough and it is easy to overcook and make it even tougher. Most of the landings, which are large, are exported or used for bait by commercial and sport fishermen. Some Canadian product may be available cleaned.

CALIFORNIA SQUID: *Loligo opalescens*. Also called Monterey squid, San Pedro squid and market squid. Huge quantities are caught, most years, when the squid enter the bays to spawn. San Pedro squid is generally small, with tubes about three to four inches long. Monterey squid is larger, with a thicker mantle, usually around seven inches long. The Monterey season in northern California is late spring and summer. In southern California, the season is December through about April. In both areas, the squid appears and disappears rapidly. Huge quantities are landed, but fishing is erratic.

California squid is white meated and tender, unless overcooked. It is generally available cleaned, representing one of the better buys available to seafood users.

STEAK A slice of fish with two parallel surfaces, subdivided as necessary. Fresh fish users should cut their own steaks from dressed fish or chunks, to avoid problems with the cut surfaces drying out.

ST. PETER'S FISH This name has been used traditionally for the John Dory, although the FDA does not now approve its use. It is also being used by tilapia marketers to entice people to try their wares. The FDA frowns rather harder on this usage. The name is not approved for any fish sold interstate in the United States.

STRIPED BASS *Morone saxatilis.* The species is generally known as rockfish in the Chesapeake region. It is a large sea bass greatly prized by sport fishermen on the Atlantic coast. When available commercially, it is a high priced fish, indicating that consumers also think it good. The flesh is coarse and rather soft, with a large flake. If it less than perfectly fresh, it can taste sour, but the fish has a mystique which ensures its marketability. Small fish are more palatable than large ones.

Striped bass are anadromous fish, with the Hudson River and Chesapeake Bay the main spawning areas. The fish is caught from Florida to Massachusetts and a few are seen in the Gulf. They used to reach 50 to 60 pounds. Striped bass have also been transplanted successfully to the west coast. No local market has developed there, but supplies can be available to meet some of the east coast demand.

The resource collapsed in the late 1970s, leading to increasing restrictions on fishing, until all fishing ceased. Because of the difficulty of policing fishing bans, some states also prohibited the importation of legally caught fish from the west coast and even of the similar farmed hybrid fish, the sunshine bass. The resource now appears to be in the recovery phase of its cycle and some very limited fishing was first allowed in 1990. Before you use striped bass, check current state regulations about ownership and fishing. These are changing, but penalties for having illegal fish can be severe.

Even if the stocks recover to abundant levels, the chances are that much of the fishing will be reserved for recreational anglers. In the past, when striped bass were important commercially, a great deal of the supply came from sport fishermen selling their catch. This fish had considerable quality problems, since anglers like to bring the whole fish ashore for weighing and photographing. Like any other fish, quality suffers badly if it is not bled, gutted and iced as soon as it is caught.

SUNSHINE BASS: *Morone chrysops x saxatilis.* This is a hybrid of the freshwater white bass and the striped bass. It looks almost identical to striped bass. It is farmed in fresh water, grows fast and has an eager market following as a replacement for striped bass. Restrictions on its possession will no doubt ease as the striped bass resource recovers. Because it is handled properly, the flesh quality and shelf life of the hybrid is much better than that of the striped bass.

STURGEON *Acipenser spp.* These fish are best known for their eggs, which become caviar. The meat of some species is also prized: English monarchs required any caught in the kingdom to be donated to the royal household. There are eight species in the United States, all of which are now scarce. Only two are used commercially.

WHITE STURGEON: *Acipenser transmontanus.* This is an anadromous species, found in rivers from Baja California, Mexico to Cook Inlet, Alaska. Fish over six feet long are now protected as they constitute the breeding population, but in earlier times the largest ever recorded weighed 1,800 pounds and was caught in British Columbia. A six foot sturgeon weighs between 80 and 120 pounds. Fish under four feet (about 20 pounds) are also protected.

Sturgeon meat is firm and well flavored and is becoming more popular as it becomes better known. The spiny buttons along the back and sides are removed and the fish is then offered headless and dressed or as fillets and steaks. White sturgeon is now being successfully farmed in California, with the ultimate objective of producing caviar, which requires large, mature fish. Meat from farmed sturgeon is available. The fish are harvested when they are about 10 pounds.

GREEN STURGEON: *Acipenser medirostris.* The green sturgeon is another anadromous fish from the west coast. The flesh is red and has a strong flavor. It is used mainly for smoking.

SUJIKO
Salted salmon eggs. This product is packed for the Japanese market. Salmon eggs are treated with salt and nitrites and then packed in wooden boxes. Sujiko is whole roe, in its skeins. Ikura is a similar product where the eggs are separate, removed from the sac.

SULFITES
Sulfiting agents such as sulphur dioxide (which is a GRAS product) are widely used in the food industry. In seafoods, they help to delay melanosis of shrimp. Because a small number of people are allergic to sulfites, current FDA standards limit the residual on the edible portion of shrimp at 100 parts per million. Use of sulfites should be noted on labels. The topic is controversial and the rules may change.

SUNFISH, FRESHWATER
Lipomas spp. and others. Freshwater sunfish, bluegills, crappie and similar fish are probably the most commonly caught game fish. All are small, edible panfish. None are used commercially any longer. Laws in many states prohibit commercial trade in these species.

SUNFISH, OCEAN
Mola mola. This is a sometimes huge fish with virtually no tail, seen lying on the surface of the sea and easily harpooned. It grows to as much as 2,000 pounds in the Atlantic, generally smaller (but still perhaps 1,100 pounds) in the Pacific. Small sunfish of around 20 pounds are often taken in gillnets.

There is no established market for sunfish. The flesh is jelly-like, but is reported to taste like lobster meat when cooked. It is eaten raw in

Japan. In Italy, the intestines are used. They are said to be similar to beef tripe.

SURIMI Surimi is a paste made from minced and washed fish flesh. The production of surimi is of major importance to the U.S. fishing industry on the west coast. The product originated in Japan. The word is used generically to describe the many products that are made from this raw material, such as imitation crab meat, imitation scallops and imitation lobster meat. Production of these, as well as of the surimi base from which they are made, is now well established in the United States, where the industry sells a good proportion of its output fresh rather than frozen. To produce surimi, sugar, salt and phosphates are added to the washed, minced flesh. The remaining salt soluble proteins, when gelled by heating, become resilient and springy and can be formed into products requiring texture. Surimi is frozen in blocks for further processing. It can be used in a variety of ways, mixed with many different colorings and cooked in several ways: steamed it is called kamaboko; broiled products are called chikuwa; and fried versions are called satsuma-age. Most products sold in the United States are kamaboko type and kamaboko has become another generic description for surimi seafoods.

Soluble proteins and most of the minerals and bacteria which produce off-flavors and off-odors are washed from the fish in the manufacture of surimi, thus preserving it. The original purpose of the product, before refrigeration, was to preserve fish for later use. The development of freezing and of factory trawlers, however, allowed huge quantities of the raw material to be made. A great deal of surimi is made aboard factory ships from Alaska pollock caught in the North Pacific.

Production of surimi is precise and exacting. Quality standards are tight (though quality means the 'elasticity' of the gel as much as it refers to freshness, flavor or other more usual features). Properly made surimi keeps for long periods in low temperature storage. Further, kamaboko products made from it also have good keeping qualities. The quality of surimi can vary widely according to differences in temperature during processing, the state of the raw material and other factors.

Surimi is frozen until required for kamaboko production, when it is thawed, chopped and mixed with salt, sugar, flavorings, colorings and various other chemicals to produce the texture, color and flavor desired.

Fresh product is generally superior to frozen for three reasons:

- Frozen surimi products tend to release moisture when thawed, which can be unsightly in use.
- Flavor and color are both lost over time and the period of transit for imported product from Japan and South Korea to the U.S. market is necessarily long, giving an advantage to domestic product.

- Finally, kamaboko for fresh distribution can be made with little or no sugar, while frozen product requires as much as five percent sugar to help preserve it. American taste seems to prefer the less sweet product and this can only be made for distribution fresh.

Surimi can be made from many different fish. Currently, Alaska pollock is the most important. Another North Pacific fish found in great abundance, Atka mackerel, can be used. So can some of the plentiful and cheap fishes found in the Gulf of Mexico.

The initial popularity of surimi products was unfortunately based on the idea that they could be substituted for much more expensive, real product. Although the FDA requires the word imitation to be used on the label, the use of imitation crabmeat in salads in restaurants is very easy. Consumers have become more sophisticated about this. Surimi products have established markets in their own right, but substitutions continue which damage the reputation of the industry as a whole and of the victimized products in particular.

Labels must state the ingredients and additives used. This is important because some people have allergies to shellfish and a smaller number have allergies to fish. Since some surimi products are made from mixtures of fish and shellfish ingredients, consumers need to know what it is they are buying. Clearly, this is not obvious from the appearance of the product.

Use of surimi is not limited to the seafood business. The raw material can be used for almost any manufactured product requiring protein. Hot dogs, pizza toppings and sausages are among other products being tested. The FDA has approved the use of surimi for some of these meat products.

SUSHI

These varied and artistic preparations based on raw seafoods have become enormously popular in the United States in recent years. Many supermarkets now offer sushi, ready to take home and eat. Consumers of sushi eat a lot of fish and shellfish and are important to the seafood business.

There are inevitably risks in eating raw seafoods. Clean, sanitary conditions during handling, distribution and preparation are absolutely essential. There is a risk of ingesting parasites, especially tapeworms and roundworms, from raw fish. Tunas, for example, which are used widely in sushi, can be heavily parasitized. Parasites, as well as many bacteria, can be destroyed when seafoods are properly frozen and stored for three or four days. Freezing greatly reduces the risks from sushi. Some more enlightened restaurants and chains require all the seafoods they use for sushi to be frozen. Fresh seafood, in peak condition, which has been frozen fast and properly stored and handled is in any case usually fresher than refrigerated seafood that has deteriorated during the time spent in transportation. For sushi, the benefits of using frozen fish and shellfish are considerably grater than a simple improvement in quality. Frozen product offers greater safety to the customer and therefore greater security to the businesses serving the customer.

SWORDFISH *Xiphias gladius.*

This is a large fish, found throughout the world's temperate seas. Commercially available swordfish in the United States are generally between 50 and 400 pounds, though a few smaller ones down to 20 pounds are also used. Only a few fish over 200 pounds are caught. Fish under 50 pounds are called pups and have a limited market, since the flesh yield is less than for large fish: it may be as little as 65 percent compared with up to 82 percent from the largest headless and dressed fish.

In the summer months, when swordfish is landed from California, the Gulf, Florida and the Atlantic coast, most swordfish is sold fresh. Quantities are also frozen for use in the winter and further supplies of frozen swordfish are imported from many countries throughout the world. The fish is the same species everywhere, so distinctions in the quality of fish from various origins are mainly due to handling and processing differences, not to intrinsic differences in the nature of the flesh. Domestic resources are under pressure.

Most swordfish is caught commercially by longline. Harpoons are now seldom used, although harpooned fish could be bled, improving flesh color and shelf life. Longline fish is often dead when brought on board the fishing vessel.

The fish is headed and gutted as soon as it is caught. The belly cavities are stuffed full of ice and the fish are buried completely in ice. Properly handled, it will keep in good condition for a couple of weeks or more. It is important that the fish be taken out of the ice every two days, washed down thoroughly inside and outside and re-iced. This ensures that the ice is always in close contact with the fish, thus keeping it at the proper temperature and preventing bacteria from multiplying.

Swordfish has very firm, solid flesh, almost the texture of pork chops. The flesh color varies considerably according to the diet of the individual fish. Some of it is pink, known and liked in New England as salmon sword. Mostly, it is off-white to pale brown.

The fish is normally sold headless and dressed, or in the form of slabs and chunks, which are fillets and cut fillets. Wheels, which are huge steaks cut through the cross-section of the fish, are also offered. The center cuts are considered preferable, because the yield is better. These may cost more. Swordfish is eaten in the form of steaks, which are produced by cutting fillets, slicing the fillets and then dividing the pieces into appropriately sized portions. Slices between three-quarters of an inch and one inch in thickness are usual. Slices should be uniform in thickness so that steaks cook in the same amount of time. Unlike steaks from most fish, swordfish steaks should be boneless; the lack of bones is one of the major attractions of the fish to consumers. The use of steaks is universal throughout the United States, as is the consumption of swordfish, which is nationally a popular high priced product.

Swordfish sometimes have very large parasitic worms, which although harmless do not look appealing. Segments of these noticed in steaks should be removed before the fish is cooked. Parasites may cause the fish to be rejected by some buyers.

More seriously, swordfish accumulate mercury from their food. Partly because the fish live a very long time, the mercury content in the flesh increases to a point where the FDA regards it as a problem. The U.S. limit for mercury is 1 part per million and interstate trade or transportation of products exceeding that level is prohibited, as is importation. Since the larger fish are older and have had longer to accumulate mercury, importers tend to bring in fish under about 120 pounds to avoid the problem: smaller fish are nearly always within the limit.

Sharks may be substituted for swordfish. Shark can be distinguished by its rough, sandpapery skin. Swordfish has a smooth skin.

Very occasionally, a swordfish with soft, jelly-like flesh is found. This condition cannot be seen until the fish is cut. It is a natural condition of the fish, cause by a microscopic parasite, which unfortunately makes the fish quite useless for consumption. A similar condition is even more rarely found in halibut. The only solution is to seek and secure credit for the affected fish.

T

TARAKIHI *Nemadactylus macropterus.* A morwong from southern oceans from South America to New Zealand. Called a jackass fish in Australia, it grows to about two pounds and has moderately firm, white flesh.

TARAMA A Greek fish roe preparation, using carp or mullet (goatfish) roe. Cod roe is also used. Salt, breadcrumbs, oil and lemon juice are added to the roe.

TAUTOG *Tautoga onitis.* Locally called blackfish, especially in New York, this is a wrasse found on the northeast coast. Most tautog are caught around three pounds, though the species will grow over 25 pounds. The flesh is white and extremely firm, with a mild flavor. It is highly regarded locally where it is caught, especially by fishermen who often keep for their own consumption the fish they find in their lobster traps.
   Supplies are limited, but the fish is very good and not expensive. The skin is tough and should not be eaten. Cunner is a similar fish, rather smaller.

THERMOGRAPH Recording thermometer. Usually a sealed device which records temperatures over a period of time. If a thermograph is carried in a truck, for example, this gives proof whether temperatures were properly maintained. For expensive and delicate shipments, it provides a simple way to check whether handling was correct. It is surprising that more use is not made of these devices, especially since temperature fluctuations in storage and transit can so severely damage the shelf life of product.

THREAD HERRING *Opisthonema oglinum.* An abundant Atlantic species. A very similar herring is found off the Pacific coast of the

United States. *Opisthonema libertate* from the coasts of Ecuador and Peru is a major resource for canned fish and fishmeal production. Thread herrings have much less oil than regular herring, so they are difficult to utilize in traditional forms. However, food technologists can design excellent product from thread herring. The large resources may offer future opportunities.

TILAPIA
Tilapia spp. Freshwater fish found worldwide. Tilapias are tough creatures and have adapted to many environments. They were introduced into the United States and in some warmer southern states have become nuisance fish, taking over habitat from more desirable species. Some states have banned tilapia. Aquaculturists have been using them to control weed in ponds and in recent years have also started to offer them as pan fish. Tilapia is now being farmed extensively, with the backing of some major corporations.

Although the flesh quality varies considerably, it is seldom better than fair. Skins are tough and the meat tends to be muddy and has a rather unpleasant texture. Nevertheless, if fed carefully and harvested at a small size, tilapia can be used.

Hybrid fish with reddish skins have been developed and offered in Florida as snapper. The relationship between tilapia and snapper is that both are fish.

TILEFISH
Lopholatilus chamaelonticeps. Also called golden tilefish. A related and very similar species, *L. princeps*, may also be called ocean whitefish. Tilefish are large, deep water fish very popular in and around New York and a few other large cities, especially Philadelphia and Chicago. Tilefish have firm flesh with a mild flavor. The skin is colorful, brownish marked with yellow spots which fade somewhat after the fish is landed. They are found in deep water off the entire east coast, in the Gulf of Mexico and as far south as Venezuela. Related species are imported from Argentina and Uruguay. They are generally marketed around six to eight pounds, but may grow as large as 35 pounds. Fresh tilefish is available headless and dressed or as fillets with the pinbones. The resource varies widely from time to time.

Tilefish keeps well under refrigeration, which adds to its popularity. Market demand continues to strengthen, as the fish is good to eat. Tile fish is worth trying for many different types of seafood operation.

TOMALLEY
The fat of the lobster, equivalent to the liver. Although excellent eating, it is spurned by most American consumers. Tomalley is the greenish mass in the head cavity of the lobster.

TOMCOD *Microgadus tomcod* is the Atlantic species. *Microgadus proximus* is the Pacific species. These are small relatives of the cod. Both Atlantic and Pacific species are available, but because it seldom exceeds one pound, the tomcod is little used commercially. If it is used, it will probably not be distinguished from regular cod. It is similar in edibility characteristics to cod.

TRIGGERFISH *Balistes spp*. Also called filefish. Highly colored reef fishes that look like they belong in an ornamental aquarium rather than on the table. Triggerfish have very tough skin, albeit highly colorful. The skin has to be removed before the fish can be filleted and it is not easy to get the skin off.

Triggerfish are very deep bodied, like pompano, and are caught in Florida and the Gulf of Mexico. The meat is very good, but most dealers consider the fish not worth the difficulty of processing. It is firm fleshed, good for poaching or in recipes designed for (genuine) turbot.

TRIPLETAIL *Lobotes surinamensis*. Atlantic and Gulf species, mainly caught by recreational fishermen. The fish grows to 30 pounds or more and is good eating. There is a related Pacific tripletail which is not common.

TRIPLOIDY A technique for altering the genetic pattern so that there are two sets of female chromosomes instead of one and three sets in all instead of two. A triploid animal will be sterile. Triploid oysters have been produced which, because they were sterile, can be marketed year round (consumers tend to dislike oysters when they are full of roe, so oysters are traditionally not marketed in the summer months containing the letter 'R'). Triploid oysters also appear to grow faster. The technique has been applied successfully to other shellfish, including species of clams and scallops.

The method of producing triploid oysters cannot guarantee that all oysters in a batch will be triploid. Some will always be regular oysters. It is not possible to determine which is which until the oyster is opened. Triploid oysters had an initial commercial success, but because buyers were uncertain which shellfish were triploid, the interest seems to have waned.

TROUT, FRESHWATER (Sea trout are covered under the heading SEATROUT.) Brook trout, brown trout, golden trout, gila trout and lake trout are a few of the trout species of sport fish, some of which are very rare. Commercially, the market is totally dominated by rainbow trout, although other trout species may be farmed and available in the future.

Note that the name trout, while in most places meaning freshwater trout (and mostly rainbow trout when the name appears on menus and in retail displays), can also mean seatrout, especially around the Gulf and in Florida. Consequently, it is not possible to define trout accurately without some modifying adjective.

RAINBOW TROUT: *Oncorhynchus mykiss.* The scientific name was changed at the beginning of 1989 from *Salmo gairdneri.* Rainbow trout was originally native to the western part of North America and the eastern part of Asia. It has been introduced to many other places and is now found in fresh water lakes and streams throughout North America and in many other parts of the world. Although customary commercial fish are under one pound, specimens over 50 pounds have been recorded by recreational fishermen. The multicolored skin, which gives the fish its rainbow name, varies greatly between different populations and there are many separate races and subspecies of rainbow trout, which account for some of the different descriptive names that are used from time to time.

Almost all U.S. rainbow trout production is farmed and almost all of it comes from Idaho. Rainbows are offered fresh or frozen, dressed or boneless and in various packs and gradings. Most are graded in two ounce steps and some are offered as straight weight sizes. Preferred sizes range between five and 10 ounces. Almost all packs are shipped with the heads on the trout.

Rainbow trout are a standard and consistent product with substantial resources behind marketing and distribution. Small, local trout farms may be able to offer special sizes, packs or even flavors, but in general there is little point in searching for such supplies: using the regular and large scale Idaho producers gives you assurance of quality and consistency that is difficult to beat.

Standard rainbow trout has a creamy, delicate flesh, with a noticeable flake and an excellent flavor. Flesh of different colors, especially salmon pinks and reds, is also available, either from specially bred hybrids or simply from feeding the fish diets designed to tint the flesh. These alternative colorations do not appear to affect the flavor or texture of the fish. The FDA does not permit these pink fleshed rainbows to be marketed under the name salmon trout, although some growers are attempting to have this ruling reversed.

The steelhead is the anadromous strain of the rainbow, found in the fish's original range, the North Pacific. It is covered under SALMON. In Europe, the anadromous rainbow is usually called salmon trout. A fast growing and red fleshed strain has been developed from a hybrid of freshwater rainbow and steelhead. This fish is now being farmed in Norway, Japan and other countries. Its fast growth rate offers the possibility of less expensive farmed fish in the future.

TROUT, FRESHWATER

BROOK TROUT: *Salvelinus fontinelis*. This is an important gamefish and a less vigorous species than rainbow trout, requiring more specialized habitats. Although not common commercially, brook trout is available from anglers and also from hatcheries which sometimes have surplus fish. Brook trout hatcheries are mainly concerned with stocking streams for anglers, rather than providing table fish. Since supply sources are not really commercial, most fish has to be purchased whole, including guts. Trout is a voracious feeder and always has food in its system and so should be gutted quickly. Trout farmers sometimes starve the fish for a day or so before harvesting to allow the gut time to be emptied. When cleaning trout of any species, make sure the belly cavity is carefully scrubbed to remove all traces of the dark red kidney material along the spine. Also take care when scrubbing not to tear or otherwise damage the belly cavity wall.

Brook trout, like rainbows, are normally eaten when under one pound, but grow to seven or eight pounds. The flesh is more delicate than rainbow trout and varies in color from white to red, according to location and feed.

BROWN TROUT: *Salmo trutta*. This game fish was introduced a century ago from Europe and is found in many streams throughout the United States. It is not normally available commercially, but when it is, the same comments apply as to brook trout, above. There is an anadromous form of the fish in European waters, which is commonly known as sea trout (but do not use this form in the United States, where it applies to drum.)

BLUE TROUT: Not a fish, but a classic method of preparing trout by dipping it in a solution containing boiling vinegar. The fish have to be very fresh so that the natural slime on the skin is still present. This turns blue and gives the dish its name.

CUTTHROAT TROUT: *Oncorhynchus clarki*, formerly *Salmo clarki*. A western freshwater trout not sold commercially any more, but highly valued by anglers. The flesh is golden or red. There is an anadromous strain which is also reserved for recreational fishermen.

GOLDEN TROUT: *Oncorhynchus aguabonita*, formerly *Salmo aguabonita*. A distinct species found only in the Sierra Nevada above 8,000 feet. Obviously, this is not a commercial product. However, the same name may be applied misleadingly and incorrectly to hybrid rainbow trout with golden skin and flesh colors.

LAKE TROUT: *Salvelinus namaycush*. A large, fat trout found in lakes in the northern United States and Canada. Siscowet and humper are either variants of the same fish or slightly different species. All tend to have a great deal of oil in the flesh, which makes them suitable mainly for smoking. Lake trout reach 100 pounds. Commercial fish are generally under 20 pounds.

TUNA Tunas are large pelagic fish of the mackerel family, found throughout the world's oceans. They are the basis of a big industry producing and selling canned tuna. Catching, processing and distributing tunas are all largely separate from much of the rest of the fishing industry. Products and aspects related to the fresh seafood trade are described in this section.

In addition to the tunas mentioned below, bonito and frigate mackerel, which are covered under BONITO, are sold as tuna from time to time. There is in practice not too much difference between a large bonito and a small skipjack tuna.

Fresh tuna, which was rarely seen on American menus a decade ago, is now widely available and highly regarded. Most of it is yellowfin. Tuna is an important part of Japanese cuisine in sashimi and sushi, both of which are popular in the United States. In general, darker meat from tuna is less delicate than lighter meat and the dark red lateral strips on most tunas are unpalatable to most American tastes and so should be removed. Most tuna is improved by bleeding it as soon as it is caught. A. J. McClane in *Encyclopedia of Fish Cookery* describes how tunas in ancient European cultures were carefully butchered, with certain cuts regarded as premium, in the same way as some pieces of beef are better than others. The art of butchering tunas into different cuts is not one that has extended to the modern United States, so users today will have to be content with steaking or filleting the fish. Fresh and frozen tuna is offered as headless, dressed trunks, as fillets and slabs and as boneless chunks.

Burnt tuna syndrome (BTS) is an occasional problem. It is characterized by pale, watery meat with a sour or bitter taste. Such meat is sometimes found in the center of the fish. BTS is a result of poor handling, especially if the fish is not cooled quickly enough after capture. BTS can be detected only after the fish is cut and quartered. If the fish is cooled to nearly freezing temperatures, it is possible to take a core sample without cutting the fish. However, the problem is not common and occurs mainly with fish caught by recreational fishermen who do not appreciate the need to cool tuna especially quickly. Tunas maintain their body temperatures above that of the surrounding water and they need to be cooled rapidly or the fish spoils. In general, it is better to buy commercially frozen tuna than to look for fresh product.

In Japan, a great deal of tuna is eaten raw. Cooked tuna, according to Spanish and Portuguese cuisine, is best brined before cooking. However, simple broiling of steaks or slabs also gives excellent results. The flesh

TUNA

color lightens considerably when cooked. Pressure cooking tuna chunks for about 20 minutes in a small amount of water makes a product similar to canned tuna, though if that is what is wanted, you will almost certainly be better off buying canned tuna directly.

Types of Tuna

Note: These are defined according to common usage in the fresh trade. The FDA has wider definitions of species to allow for the worldwide sources of raw material for the canning industry. For details, see Title 21 of the Code of Federal Regulations, section 161.190.

ALBACORE: *Thunnus alalunga.* Worldwide in distribution, this is the species called white meat tuna by canners and the only species which may be so labeled. It is caught far out on the oceans and frozen on board the fishing vessels. Albacore grow to almost 90 pounds, averaging perhaps 20 pounds commercially. Skinless, boneless loins and steaks are readily available frozen. Albacore that has not been frozen is rare. Line caught fish is better than netted fish, because it is not bruised and will normally have been bled. Most albacore is brine frozen, which adds salt to the meat, but blast frozen or even plate frozen product is also available.

Albacore can be used as a substitute for chicken and veal, though it toughens if even slightly overcooked. Frozen product is usually excellent. Fresh product may have been kept in ice for too long. The fish seldom comes close enough to shore to justify it being brought back unfrozen.

BIGEYE: *Thunnus obesus.* Also worldwide in distribution, the bigeye tunas grow to 300 to 400 pounds and are an important resource of the international tuna fishery. Bigeye is sold canned as light meat tuna, the meat being darker than albacore but lighter than skipjack. Almost all the production is canned. Bigeye tuna is little used in fresh form.

BLACKFIN: *Thunnus atlanticus.* A small tuna, generally around 10 pounds but reaching 40 pounds, caught in the warmer Atlantic, especially around Florida. The meat is dark but firm and well flavored, excellent broiled. Fresh, the fish is usually sold headless and dressed. Steaks or slabs can be cut for use as required.

BLUEFIN: *Thunnus thynnus.* Bluefin is sometimes confusingly called horse mackerel, which it is not. Northern bluefin tuna is one of the largest of all fishes, reaching over 1,400 pounds. Although this fish is found in many parts of the world's oceans, product caught in the United States comes from the Gulf of Mexico and the Atlantic. It is uncommon in the eastern Pacific. In general, the fish swim north along the Atlantic coast each summer. The giant bluefin appear about midsummer off the middle Atlantic states and migrate northwards. Some of the very largest fish are regularly caught in the Canadian Maritimes. There are complex regulations governing how many bluefin may be caught each year in U.S. waters.

The meat is dark, almost red. There is a very dark red strip of meat along the lateral line, which is particularly prized by Japanese consumers. Only very small bluefin are used for canning, as part of the light meat pack. Cooked bluefin has rather firm texture and a distinctively strong flavor, which is reduced if the meat is brined overnight before cooking. The skin should be removed before bluefin tuna is eaten and it is generally removed before cooking.

Small fresh bluefin (the description small is strictly a relative term, the fish are so large) are used seasonally in east coast markets, especially in New York and Philadelphia. The fish are sold headless and dressed and require very thorough and careful icing, as well as gentle handling, if quality is to be maintained. As the fish caught get larger later in the year, around August and September, the market emphasis switches to Tokyo. Giant bluefin are airfreighted individually to Japan and sell for high and widely fluctuating prices.

Handling of giant bluefin for Japan is a specialized and skilled business. Not only must the fish be well iced and treated with the utmost care, but since the tuna is eaten raw, bruising has to be avoided. The delicate flesh can be marred simply from the heat of a hand placed against the carcass, so bluefin are moved around in slings with special gear. If it is necessary to touch them, workers wear gloves. With the growing popularity of Japanese cuisine in the United States, a market has developed for fresh bluefin— and for other tunas—produced to the same high standards. Note that Japanese fishermen catch bluefin in the Gulf of Mexico as well as in other oceans. These are frozen to especially low temperatures (below -40°F) and, like the expensively handled fresh tuna, sold for sashimi in Japan. Special equipment is needed to handle these huge fish at such low temperatures.

LITTLE TUNA: *Euthynnus alletteratus*. An Atlantic species, similar to the blackfin tuna, normally around five pounds and reaching perhaps 15 pounds. In 1990 FDA approved the common name of spotted tunny. This species is frequently the tunny fish found in Mediterranean recipes. The flesh is dark and strongly flavored, best when brined before cooking. It also makes excellent smoked products.

SKIPJACK: *Euthynnus pelamis* is the scientific name preferred by the American Fisheries Society. However, *Katsuwonus pelamis* is still widely used. There are substantial scientific differences of opinion about which is the definitive name for this species. Do not confuse the name skipjack with small bluefish, which are also called skipjacks. Skipjack is a small tuna, generally around seven to 12 pounds, that's found worldwide and is an important fishery through the central Pacific (it is known as aku in Hawaii, which is the major domestic source of frozen skipjack). Skipjack has fairly dark meat and is a major part of the light meat canned pack. It is occasionally available frozen, seldom fresh, since it has a rather short shelf life. It is palatable broiled. Skipjack can replace bonito or horse mackerel in most recipes. Tuna fishermen tend to look only to the canneries for their market, but skipjack could probably find more frozen markets domestically.

YELLOWFIN: *Thunnus albacares.* Yellowfin is the major catch of the California tuna fleet and is also found in the Gulf Stream waters of the Atlantic, providing an important harvest for Florida fishermen. Generally between 20 and 120 pounds, fish as large as 388 pounds have been caught. Yellowfin is light meat tuna. The meat is lighter than that of skipjack, darker than albacore.

Known as ahi in Hawaii, yellowfin has developed a good market following in recent years for fresh and frozen product. It must be bled on capture and rapidly iced, but fresh product of adequate standard is available.

TURBOT *Psetta maxima.* Do not confuse this fish with greenland turbot (see next heading). Genuine turbot is a large flatfish found in the North Sea and eastern Atlantic, but not on the U.S. side of the ocean. It has firm, sweet flesh and is one of the outstanding flatfish to eat. High priced and not common, turbot is occasionally imported fresh, but more often frozen, headless and dressed, graded in one pound steps from three to 12 pounds. Fish between six and nine pounds are preferred. Turbot is usually steaked rather than filleted for serving. Turbot is now being farmed on a small scale in France and Spain. The technology is being developed in other counties also. It is possible that, if strong demand for this expensive fish can be met in Europe, there will be surplus quantities available in the future for the U.S. market.

TURBOT, GREENLAND *Reinhardtius hippoglossoides.* This fish may legally be called greenland turbot and greenland halibut. In Europe, it is known as black halibut or mock halibut. This is a large flatfish, but it has moderately soft, rather tasteless flesh. Huge quantities of this species and some related Pacific species are imported as frozen skinless fillets. Some of these are repackaged and appear on the market as flounder or sole. It is cheaper than both and inferior to both. Canadian origin fish from the east coast may be offered fresh, when it can be used as an inferior and usually large flounder (16 ounce fillets are common).

West coast Canadian turbot may be offered. This is actually arrowtooth flounder, it is very soft and tasteless and tends to collapse into paste when cooked because of the presence of a microscopic parasite. Although the turbot designation appears to be correct in Canada, it is not turbot in the United States. This particular fish should be avoided, however cheap.

U

ULUA Hawaiian name for several species of jacks, which grow as large as 150 pounds. These fish have pink flesh, which turns white when cooked. They are highly regarded in Hawaii, where they are used for sashimi as well as regular cooked dishes.

UNDERUTILIZED SPECIES There are many seafood products less costly than the standard favorites. There are those used as substitutes for better fish and shellfish (see SUBSTITUTIONS). There are others which are described as underutilized, meaning there are large supplies available in the seas but little market for them. Some of these are cheaper because some particular feature may be less desirable, not suitable for processing machinery or in some other way be different. There is a place for many of these products if the buyer accepts that they must be used in ways which utilize the intrinsic features of the product, not necessarily in ways which conform to current preparation and cooking habits.

The United States is especially well provided with marine resources which are little used and offer scope for exploitation and development. The question is how to take advantage of this for individual and national benefit.

The potential is hard to define exactly but clearly very large. Scientists estimate that perhaps 250,000 tons of blue runner could be landed in the southeast. That is about 500 million pounds a year. Atka mackerel from Alaska, red hake, squid, dogfish and many others are available in varying and substantial quantities but are caught in only small amounts because there is no market for the species. There have been some notable successes in developing industries on new resources. Alaska king crab was once an underutilized resource, strange as that now seems. Squid exports from New England are now substantial and even U.S. consumers are finally showing an interest in squid rings.

American consumers enabled New Zealand to exploit their orange roughy. But there are plenty of opportunities still remaining.

The role of seafood distributors, restaurants and retailers is crucial. Consumers cannot decide a product is acceptable unless they have the opportunity to try it. Too often, the trade decides that a product will not sell, refuses to handle it and proves itself right: it does not sell because no one is willing to try to sell it. This does not mean that you should buy every unusual item that is offered to you. It does mean that you should examine the potential they may have for you.

Successful use of an underutilized species offers lower priced raw material, which helps you to compete on price. It also helps you generate interest and excitement, which are the best marketing tools available to the seafood industry.

The critical elements in marketing underutilized species are as follows.

NATURE OF THE PRODUCT: This is a question initially of individual taste and judgement. You have to decide that the product is palatable, attractive and saleable. Unfortunately, the judgement of the public frequently differs from the opinions of people in the trade. Throughout the food industry, many new products, often launched after extensive and expensive research into their acceptability, fail miserably—overall over 90 percent of new consumer products fail even to reach national distribution. Since there is little you can do to insure yourself against this sort of problem, the best advice is to trust your own taste and instincts. If you think the product is good and that it fits with your present product lines, it is worth trying. If you think the product unacceptable, you should not inflict it on your customers. Nor will you have sufficient enthusiasm to push it properly.

WHAT CAN YOU CALL IT? This is perhaps the most critical problem in marketing underutilized species. The name king crab is clearly more attractive than spider crab, which was a possible name for the product. Squid has negative connotations for many consumers who may be tempted by calamari or by golden rings or similar euphemisms. Dogfish will not sell as well as rock salmon, which is what dogfish used to be called in Britain.

You cannot simply pick an attractive name without regard to the species and customary name of the fish. The FDA regulates the use of names for seafoods. The agency has developed a procedure for selecting names for new species. See the heading NOMENCLATURE for details. In general, the rules are designed to protect consumers from being sold a product other than the one they expected to purchase. To see if consumers find the fish acceptable at the restaurant level, fish and chips,

UNDERUTILIZED SPECIES

catch of the day and similar generic terms can be used to experiment with the acceptability of unusual fishes.

ASSURANCE OF SUPPLY: Effort applied to helping a supplier dispose of an accidental load of an unusual fish is wasted. Before you agree to buy any unusual item, make sure that you have some guarantees of continued supply if your customers like the product. Also make sure that such supplies will be at or near the price paid for the original shipment. If an item is being caught and processed for regular export sale, you have some assurance that supplies will be available since production will be less speculative for both fishermen and processors. It is also more likely that you will find alternative suppliers. In such circumstances, your demand will be competing with the export market demand, which should also be checked carefully.

There is no supply or price assurance for any natural seafood product. Nevertheless, you need as strong an assurance as you can reasonably get to make it worth experimenting with underutilized species.

V

VEIN The vein or sandvein is the lower intestinal track of a shrimp, lobster or similar crustacean. This runs through the tail of the animal and looks like a thin black or brown tube. It is called a sandvein because the contents are frequently gritty, like sand. The vein varies considerably in different species. In some, it is quite prominent. In others, it is pale and hardly noticeable. Freshwater crayfish may be purged before sale to make the vein barely noticeable.

Although the vein is not harmful to eat when cooked, it is aesthetically undesirable and therefore frequently removed, either before preparation or by the diner.

VENDACE *Coregonus albula.* A European freshwater fish related to whitefish. The roe is used for a caviar substitute in Sweden.

W

WAHOO *Acanthocybium solanderi.* A large relative of the mackerels, reaching over 100 pounds and important as a game fish off southern Florida. It is appreciated in Hawaii, where it is called ono. It is a good food fish, similar to the large mackerels.

WATERMARK As salmon swim upstream to their spawning grounds, the skin colors change dramatically. The original silver fades. Then reddish, gray and black patches and streaks begin to appear. Some salmon finish up with almost black skin, others with almost red skin. Watermarking is actually an indication of the sexual maturity of the salmon, which changes shape as well as color as it matures and reaches the spawning grounds.

Watermarks begin high on the back of the fish and spread down towards the belly. The quality grades of various species of salmon change as watermarks develop. These are defined and discussed under SALMON.

WEIGHTS AND LABELING In principle, net weight is a simple concept. Net weight is the weight of the commodity without any packaging and without other materials packaged with the commodity. In other words, it is the weight of the product without glaze, without added water or broth and without packing material. Most state rules allow small variations between individual packages, provided that the total weight of a shipment is correct. Since shipments are split up, it is not safe to rely on this too heavily. States legislate for trade within their borders, but most have adopted uniform rules. Federal law supercedes state law for interstate shipments (including imports).

Although the principle of net weight is clear, it can be hard in practice to determine how much product was originally shipped. Fresh seafoods naturally drip: soluble proteins slowly seep out of the product. Some seafoods, scallops especially, drip rapidly and can lose an alarming proportion of their weight. This is one reason for using dips which help to retain moisture (see DIPS). Frozen seafoods, which must be protected with an ice glaze, are more vulnerable to cheating on the weight, since determining the unglazed product weight is not always easy and often requires properly controlled laboratory procedures.

The National Conference on Weights and Measures has a guideline for selling shellfish, as follows:

- Stuffed clams or mussels on the half shell should be sold by net weight, excluding the weight of the shell.
- Canned oysters should be sold by net weight.
- Fresh, shucked oysters, clams or mussels should be sold by fluid volume.
- Frozen oysters, clams or mussels should be sold by net weight.
- Whole clams, oysters or mussels in the shell (fresh or frozen) should be sold by dry measure (e.g., bushel) or count plus size, not by weight.

These guides are perhaps less than perfect, but a step towards achieving uniformity.

Labeling laws throughout the United States require packages to be marked with their net weight. Some states specify the position and size of the label and the printing. Others do not. The majority follow the guidelines of the *Model State Regulations* published by the National Bureau of Standards, which include suggested packaging and labeling laws and weights and measures laws. Federal law was consolidated in the Fair Packaging and Labeling Act. Labels generally have to include lists of ingredients. This means that if tripolyphosphate or other dips are used, they must be listed on the label.

WHELK
Busycon carica is the knobbed whelk. *Busycotypus canaliculatus* is the channeled whelk. Although these are the official English names preferred by the American Fisheries Society, the species are more commonly called snails, conch or scungili. Although the last is actually a particular way of preparing the food, it is the only name commonly used which does not promote confusion with some other type of animal.

Scungili, or whatever they should be called, are large, single shelled marine snails growing to as much as nine inches in length. They are found from Cape Cod to Florida. Most of the commercial catch comes from Massachusetts and Rhode Island, though there is a consistent, but small, fishery in South Carolina. Live whelks survive for months out of water if kept damp and cool. Most end users prefer to buy cooked meats, which are usually offered as "fully cooked" but are in practice usually cooked only long enough to facilitate removing the meat from the shell.

Although the meats can be removed without cooking by cracking open the shell, raw meats have a very short shelf life, turning sour almost immediately. They should always be frozen.

It is generally preferable to buy cooked conch meat. Because commercially cooked conch may have been processed to make it easier to remove the meat from the shell rather than to prepare it for eating, it is important to check that the meat is fully cooked. Otherwise, you have to cook it all over again, with the associated smell and the shrinkage that results from re-cooking the shellfish. Most processors claim that their conch is fully cooked, often that theirs is the only one so processed. In practice, not much conch is fully cooked. Conch is one of the few, probably the only, shellfish that requires extended cooking. Most shellfish toughen if cooked for more than a few minutes, while conch requires sometimes as long as an hour, and softens as cooking time is lengthened.

Conch is frequently made into "salad," using acidic marinades. The marinade will soften the meat sufficiently, even if it was only partially cooked to begin with.

The processor will cook live conch, remove it from the shell and discard the foot and attached viscera. Some will remove the rest of the viscera, some will not. These remaining internal parts are edible, but have a spongy texture which contrasts oddly with the chewy meat.

Cooked conch turns sour quickly. It must be kept well iced and refrigerated. For most purposes, it is better to use frozen conch rather than risk losing fresh product. Freezing does not impair the taste or texture at all. Frozen conch is often packed in vacuum bags which help to delay the onset of rancidity.

WHITEBAIT Young or very small herrings, sardines, sandeels, silversides or similar fish. Whitebait are fried and eaten whole in Europe, but are little used in the United States. There is a whitebait smelt that is caught on the west coast, but this is used mainly for bait.

Whitebait is an attractive and inexpensive dish, if you can persuade suppliers to ship the small fish, fresh or frozen, in prime condition. Because the fish are whole, feed in the gut ferments, so they must be kept very cold and used quickly. Whitebait ideally should be less than three inches long.

WHITE BASS *Morone chrysops*. A freshwater species, generally around two pounds, with good white flesh. Commercial fishing in the Great Lakes is supplemented with a large recreational catch throughout the central and southern portions of the United States. White bass are usually offered dressed. A hybrid of this species and the anadromous striped bass is now being farmed successfully. See STRIPED BASS.

WHITEFISH A general term covering fish with white flesh, such as cod and flounder, differentiating them from fatter fish such as herring and mackerel, which have darker flesh. The term is imprecise and confusing and should be avoided. For example, sablefish has white flesh but is an oily fish, so it is not clear whether it is a whitefish.

WHITEFISH, LAKE *Coregonus clupeaformis*. Larger than the cisco (about two to three pounds, although anglers once claimed 20 pound fish), whitefish are generally sold smoked in the United States. Like cisco, they have well flavored, white flesh and can be used for most fish recipes. Catches have declined and most supplies now come from Canada.

WHITING Hake is also covered in this section. There are scientific distinctions between hake and whiting, but the commercial rule, for the U.S. market, is that whiting is a better market name than hake because consumers prefer to buy and eat whiting. Pacific hake legally became whiting in 1979, when the FDA agreed to the name change. Hake and whiting are related to cod. They generally have longer, thinner bodies.

Most people regard hake as a large, soft fleshed fish. White, silver and red (or squirrel) hake are all available types. Whiting is generally regarded as a smaller fish, firmer than hake. A lot of whiting and hake comes from South America. To add to the confusion, the Spanish word for both whiting and hake is *merluza*.

Hake and whiting have soft flesh, with little flavor. If it is accepted that the distinction is that hakes are large and whiting are comparatively smaller, then hake is best used in steak form, since the bones in the center of the steak help to prevent the fish from falling apart. It may also be used headless and dressed for stuffing and baking. Whiting are most frequently sold headless and dressed, in which form they are fried. A great deal of whiting is smoked, but smokers prefer imported product from South America. Domestic whiting from the West Coast varies from difficult to impossible to smoke, because of enzymes which are naturally present in the flesh, and cause the flesh to soften and break up when the fish is smoked.

Comparatively little whiting is sold fresh, partly because of the short seasons and partly because of the abundant and cheap supplies of imported, frozen fish, available headless and dressed or as fillets and skinless fillets. Whiting is difficult to handle fresh, as it goes stale very quickly. Pacific whiting, in particular, requires freezing as soon as it is caught. The flesh breaks down naturally at high speed (again, because of the enzyme which is produced by a myxosporidian parasite), making handling this species fresh extremely risky.

WHITING

There are a number of whiting species imported in large quantity from Peru, Argentina and other countries. These are invariably frozen product, inexpensive and an excellent value in terms of protein for the dollar. Commercial hakes and whitings that might be offered fresh include the following.

ATLANTIC WHITING: *Merluccius bilinearis*. This fish is also legally called silver hake, although whiting is preferred for marketing. It is similar to red hake. Whiting is generally about 12 ounces, although the species may grow as large as five pounds. Processed, the headless, dressed fish are usually under eight ounces and the fillets under three ounces. Larger whiting are sometimes mislabeled as king whiting.

In some years, enormous quantities are landed in short seasons in the northeast. Most of the fish is frozen in headless, dressed form, packed in three pound or five pound boxes for supermarket sale. Fresh whiting is also a popular supermarket item when it is available cheaply during the season. July and August are the peak supply months, with lower landings made until November. However, supplies vary enormously from year to year.

ANTARCTIC QUEEN: *Merluccius australis*. This species may also be called whiting and New Zealand whiting. In New Zealand, it is generally known as hake. Do not confuse it with hoki, *Macruronus novaezelandiae*, which is also allowed to use that name but is generally a rather softer fish. The FDA approved the name antarctic queen in 1988.

The species is quite large, with fish of eight pounds not uncommon. It has very white flesh, firmer than most other whiting. Headless and dressed fish is sometimes available fresh from Chile, though it is debatable whether it is worth the high price required to repay air transport costs.

PACIFIC WHITING: *Merluccius productus*. FDA rules permit this fish to be called whiting, Pacific whiting or Pacific hake. There is a major resource of this species off the coasts of Washington and Oregon. Factory trawlers are starting to exploit it. There is a much smaller resource in Puget Sound, Washington. This has been exploited for many years. There is a distinction between the Puget Sound and the ocean fish, although they are the same species. Oceanic fish are larger at 19 to 26 inches, compared with 14 to 18 inches for the Sound fish.

A substantial proportion of Pacific whiting carry a microscopic parasite which, when the fish is caught, produces an enzyme that breaks down the flesh. If the fish is frozen very soon after capture, the enzymatic action is virtually halted. However, the fish must either be cooked from the frozen state or thawed very rapidly immediately before cooking, otherwise the ing, otherwise the problem reappears. This species is not suitable for handling fresh, except within a few hours of the fishing grounds. The parasite also makes it unsuitable for smoking.

RED HAKE: *Urophycis chuss*. Sometimes called squirrel hake, though the name is not approved by the FDA. Red hake is a North Atlantic species, abundant from Maine to New Jersey. Most commercially available fish are about one pound. Spawning fish in early summer have especially soft flesh. At other times, the flesh has a delicate flavor but needs careful handling. The resource is much larger than the present market and could be substantially developed, if there were a market. Much of the red hake now caught is discarded by fishermen.

WHITE HAKE: *Urophycis tenuis*. Generally one to two pounds, this North Atlantic species is sometimes very much larger, up to 25 pounds. In Boston and in Maine, it is usually landed headless and dressed and is used for salting or for steaks.

WHITING, BLUE *Micromesistius poutassou*.

An abundant deep water relative of the whiting, fished in the middle of the Atlantic by factory vessels equipped for making this small, soft fish into blocks. There seems little chance that this fish would be offered fresh. A similar species, the southern blue whiting (*Micromesistius australis*) also appears to be abundant, though resources around the Falkland Islands are reported to be heavily parasitized, making the fish at present suitable only for fishmeal.

WINKLE *Littorina littorea*.

Also called periwinkle and sea snail. Winkles are small marine snails found in large numbers along the shoreline of the northern Atlantic. Only very small quantities are used in the United States. The snails are shipped alive in bushel bags. They are boiled alive and eaten straight from the shell. The operculum, which is the plate covering the opening of the single shell, is removed; the rest of the animal can then be eaten. In England and France, the meat is traditionally extracted by the diner with a pin, preferably gold plated.

Winkles are abundant. They can be seen in huge numbers around the shoreline, especially on rocks. The problem is harvesting them. Winkles must be collected by hand. This is boring and onerous work. It is also backbreaking.

It is important to cook only live winkles, since a couple of dead ones can give a very bad flavor to all the others cooked with them. Winkles should be washed and then cooked for about 10 minutes in boiling, salted water. In France, they are often served to diners waiting for oysters to be opened.

Periwinkle is probably the most used name for these little snails, but the same word is used regionally for the northern conch or whelk.

WRECKFISH *Polyprion spp.* A large sea bass caught off the southern Atlantic states. It grows to a length of about five feet and a weight of 100 pounds. A fishery grew rapidly in the late 1980s, but the resource collapsed. Catches are now strictly limited. The fish is managed with other reef fish (snappers and groupers). Commercially, wreckfish is very similar to grouper and is treated as if it were grouper.

Y

YELLOWTAIL *Seriola lalandei*. A large member of the jack family, caught off California and Baja California, mainly by recreational fishermen. It is similar to the Atlantic amberjack, which is sometimes substituted for it. Yellowtail are generally around 30 pounds, though the record specimen weighed 80 pounds. Headless, dressed fish and fillets are sometimes available. It has good flavor and the flesh is not as dark as that of amberjack.

Japanese yellowtail, *Seriola quinqueradiata*, is a different and much more expensive fish which has been farmed in Japan's Inland Sea for many years. It is delicate and oily, used raw for sashimi. Japanese yellowtail is flown into U.S. markets, mainly California, for use in sushi restaurants. The domestic yellowtail is not considered equivalent.

The word yellowtail also refers to a FLOUNDER and to a SNAPPER. For information, check those headings.

YIELD Yield is the percentage of a fish or shellfish that is edible (the edible yield) or saleable (the market yield). Appreciating yield is crucial to successful buying, especially if you have the option of cutting whole fish or buying fillets.

Cod, pollock, or similar fish, medium size and in good condition (meaning the flesh is firm, the fish is freshly caught and it has been well iced) will yield a good cutter about 43 percent of the gutted weight in the form of skinless fillets with the pinbones in. With the skin left on, the yield would be about 46 percent. Less skillful cutting, softer fish, smaller fish, fish with roe, machine processing and many other factors can lead to lower yields.

On this general basis, however, a fish landed head on and gutted at $1.00 per pound would give a fillet cost for raw material of $1.00 divided by 43 percent, which is $2.33 per pound. To that must be added the costs of the filleting operation, which might be 25 to 50 cents per pound (but which vary widely).

YIELD

Fish landed whole, with viscera, obviously yield lower percentages of edible fillets. In the case of cod, the yield would be about 36 percent. Other fish give quite different yields—there is no norm or standard. Ocean perch, for example, will yield only 30 percent of skin on fillets from gutted fish. Flounder and sole will give, on average, about 35 percent for skinless fillets, with extremely wide variations according to the size, species and condition of the fish.

Yield of clam meat from live quahogs should be about five quarts of meats from one bushel of large clams. Oysters should yield about four quarts (one gallon) of meats from a bushel. These yields can vary substantially.

Crabs give very small yields of meat for a great deal of labor. Even king crab only gives about 18 to 21 percent from the whole, cooked crab. Small crabs, such as blue crab, can yield even lower percentages. Live or weak lobsters cooked for the extraction of meat yield as little as 12 to 15 percent of their weight as cooked meat.

Gulf and other warm water shrimp are generally landed snapped, that is shell-on tails ready for freezing. In considering whole shrimp, which is sometimes offered, remember that the edible tail is only about one-third of the weight of the whole shrimp. Similarly, northern shrimp may be offered whole (raw or cooked) or cooked and peeled, ready for use. The yield of cooked and peeled shrimp from whole, raw is in the region of 15 percent, so without taking into account the considerable labor cost of cooking and peeling, the raw material would have to cost less than one-sixth the price of the cooked and peeled product to begin to make economic sense.

It must be stressed that these example yield figures are very approximate. There are enormous differences according to the condition of the product, the size of the fish or shellfish, the skill of the operatives and many, many other factors.

Information on yields is spread over a great deal of scientific literature. It is often difficult to find source material. *Seafood Leader* magazine offers yield data for the species it covers in its annual buyer's guide. Alaska Sea Grant has published a booklet called *Recoveries and Yields from Pacific Fish and Shellfish,* which covers over 60 species. This is extremely useful and can be used as a guide for many related species from other waters. Ask for Marine Advisory Bulletin No. 37, December 1988.

The final yield is on the plate. Fish and shellfish shrink when they are overcooked. Portions look bigger (indeed, they are bigger) if seafoods are cooked correctly.

Resources

This section lists magazines, newsletters and other publications that contain helpful information and news on seafood.

Periodicals

American Seafood Institute Report. 406A Main Street, Wakefield, RI 02879

Aquaculture Magazine. P.O. Box 2329, Asheville, NC 28802

Australian Fisheries. Department of Primary Industry, Canberra, ACT 2600, Australia

Canadian Aquaculture. 4611 William Head Road, Vancouver, B.C. Canada, V8X 3W9

Commercial Fisheries News. P.O. Box 37, Stonington, ME 04681

FAO Infofish Marketing Digest. P.O. Box 10899, Kuala Lumpur 01-02, Malaysia

FDA Consumer. 5600 Fisher's Lane, Rockville, MD 20857

Fisheries. 5410 Grosvenor Lane, Bethesda, MD 20814

Fisheries Product News. P.O. Box 37, Stonington, ME 04681

Marine Fisheries Review. 7600 Sands Point Way NE, Seattle WA 98115

National Fisherman. 120 Tillson Ave, Rockland, ME 04841

RESOURCES

Pacific Fishing. 1515 N.W. 51st St., Seattle, WA 98107

Quick Frozen Foods. 7500 Old Oak Boulevard, Cleveland, OH 44130

Quick Frozen Foods International. 80 Eighth Ave, New York, NY 10011

Seafood Business. 120 Tillson Ave, Rockland, ME 04841

Seafood International. 81-89 Farringdon Lane, London, EC1M 3LL, England

Seafood Leader. 1115 NW 46th St., Seattle, WA 98107

Seafood Supplier. 120 Tillson Ave, Rockland, ME 04841

World Aquaculture. 16, East Fraternity Lane, Louisiana State University, Baton Rouge, LA 70803

Newsletters

Aquaculture Association of Canada Bulletin. P.O. Box 1987, St. Andrews, N.B., Canada E0G 2XO

Aquafarm Letter. 3400 Neyrey Dr., Metairie, LA 70002

Cameron's Foodservice Promotions Reporter. P.O. Box 1160, Williamsville, NY 14221

Commercial Fishing Newsletter. VIMS, Gloucester Point, VA 23062

Erkins Seafood Letter. P.O. Box 108, Bliss, ID 83314

LMR Shrimp Market Report. LMR Fisheries Research Inc., 11855 Sorrento Valley Rd., Suite A, San Diego, CA 92121

Seafood Price-Current. P.O. Box 389, Toms River, NJ 08753

Seafood Trend. 8227 Ashworth Ave N., Seattle, WA 98103

Sea Grant Abstracts. P.O. Box 125, Woods Hole, MA 02543

Shrimp Notes: A Market News Analysis. Shrimp Notes Inc., 417 Eliza St., New Orleans, LA 70114

Bibliography

1953. *La Mer.* France: Librairie Larousse.

1978. *Fish Inspection Regulations of British Columbia.* Victoria, British Columbia: Government of British Columbia.

1980. *Official Methods of Analysis of the Association of Official Analytical Chemists,* ed. William Horwitz. Washington, DC: Association of Official Analytical Chemists.

1981. *Guide Book to New Zealand Commercial Fish Species.* Wellington, New Zealand: New Zealand Fishing Industry Board.

1982–1990. *Fisheries of the United States.* Washington, DC: U.S. Department of Commerce.

1984. *Manuals of Food Quality Control 5—Food Inspection.* Italy: Food and Agriculture Organization of the United Nations.

1985. *Federal, Food, Drug and Cosmetic Act, As Amended.* Washington, DC: U.S. Government Printing Office.

1985. *Meat and Poultry Inspection—The Scientific Basis of the Nation's Program.* Washington, DC: National Academy Press.

1988. *The Fish List:—FDA Guide to Acceptable Market Names for Food Fish Sold in Interstate Commerce.* Washington, DC: U.S. Government Printing Office.

1988. *Recommended Marketing Names for Fish.* Canberra, Australia: Australian Government Publishing Service.

BIBLIOGRAPHY

1988. *Seafood Safety: seriousness of problems and efforts to protect consumers.* Report to the Chairman, Subcommittee on Commerce, Consumer and Monetary Affairs, Committee on Government Operations, House of Representatives. Washington, DC: United States General Accounting Office.

1989. *Assessing Human Health Risks from Chemically Contaminated Fish and Shellfish: A Guidance Manual.* Washington, DC: United Sates Environmental Protection Agency.

1989 Revision. *National Shellfish Sanitation Program, Manual of Operations Part I, Sanitation of Shellfish Growing Areas.* Washington, DC: U.S. Department of Health and Human Services, Public Health Service, Food and Drug Administration.

1989 Revision. *National Shellfish Sanitation Program, Manual of Operations Part II, Sanitation of the Harvesting, Processing and Distribution of Shellfish.* Washington, DC: U.S. Department of Health and Human Services, Public Health Service, Food and Drug Administration.

1990. *Code of Federal Regulations 21: Food and Drugs, Parts 100 to 169.* Washington, DC: US Government Printing Office.

1990. *Code of Federal Regulations 50: Wildlife and Fisheries.* Washington, DC: U.S. Government Printing Office.

1990. *Commercial Fishing Guide for Shellfish and Minor Finfish Species.* Vancouver, British Columbia: Department of Fisheries and Oceans—Pacific Region.

1990. *Interstate Certified Shellfish Shippers List.* Washington, DC: Department of Health and Human Services.

Ade, Robin. 1989. The Trout and Salmon Handbook. New York: Facts on File Inc.

Allen, Standish K. Jr., Sandra L. Downing, and Kenneth K. Chew. 1989. *Hatchery Manual for Producing Triploid Oysters.* Seattle, Washington: University of Washington Press.

Bigelow, Henry B. and William C. Schroeder. 1964. *Fishes of the Gulf of Maine.* Cambridge, Massachusetts: Museum of Comparative Zoology, Harvard University.

Blaufarb, G.P. and E.C. Johnston. 1987. Voluntary U.S. standards for grades of fishery products. In *Seafood Quality Determination*, ed. Donald E. Kramer and John Liston pp. 665–670. New York: Elsevier Science Publishing Company Inc.

Browning, Robert J. 1980. *Fisheries of the North Pacific*. Anchorage, Alaska: Alaska Northwest Publishing Company.

Bryan, F.L. 1987. Seafood-transmitted infections and intoxications in recent years. In Seafood Quality Determination, ed. Donald E. Kramer and John Liston, pp. 319–337. New York: Elsevier Science Publishing Company Inc.

Butler, T. H. 1988. Dungeness crab. *Underwater World*. Ottawa, Canada: Department of Fisheries and Oceans.

Canzonier, W.J. 1988. Public health component of bivalve shellfish production and marketing. *Journal of Shellfish Research* 7(2):261–266.

Castro, José 1983. *The Sharks of North American Waters*. College Station, Texas: Texas A & M University Press.

Cheney, Daniel P. and Thomas F. Mumford, Jr. 1986. *Shellfish & Seaweed Harvests of Puget Sound*. Seattle, Washington: Washington Sea Grant Program, University of Washington.

Cobo, Mario and Sheyla Massay. 1989. *Lista de los Peces Marinos del Ecuador*. Guayaquil, Ecuador: Instituto Nacional de Pesca del Ecuador.

Connell, J. J. 1990. *Control of Fish Quality*. Oxford, England: Fishing News Books.

Couturier, Cyr. 1990. Scallop culture in Canada. *World Aquaculture* 21(2):54–62.

Crapo, Chuck, Brian Paust and Jerry Babbitt. 1988. *Recoveries and Yields from Pacific Fish and Shellfish*. Fairbanks, Alaska: Alaska Sea Grant College Program, University of Alaska.

Davidson, Alan. 1977. *Seafood of South-East Asia*. Singapore: Federal Publications (S) Pte Ltd.

Davidson, Alan. 1981. *Mediterranean Seafood*. Baton Rouge, Louisiana: Louisiana State University Press.

Davidson, Alan. 1980. *North Atlantic Seafood*. New York: The Viking Press.

BIBLIOGRAPHY

Davidson, Alan. 1989. *Seafood: A Connoisseur's Guide and Cookbook.* New York: Simon and Schuster.

Doré, Ian and Claus Frimodt. 1987. *An Illustrated Guide to Shrimp of the World.* New York: Van Nostrand Reinhold/Osprey Books.

Doré, Ian. 1990. *Making the Most of Your Catch—An Angler's Guide.* New York: Van Nostrand Reinhold/Osprey Books.

Doré, Ian. 1989. *The New Frozen Seafood Handbook—A Complete Reference for the Seafood Business.* New York: Van Nostrand Reinhold/Osprey Books.

Doré, Ian. 1990. *Salmon—The Illustrated Handbook for Commercial Users.* New York: Van Nostrand Reinhold/Osprey Books.

Doré, Ian. 1991. *Shellfish—A Guide to Oysters, Mussels, Scallops, Clams and Similar Products for the Commercial User.* New York: Van Nostrand Reinhold/Osprey Books.

Doré, Ian. 1991. *Fish and Seafood Quality Assessment: A Guide for Retailers and Restaurants.*

Faria, Susan M. 1984. *The Northeast Seafood Book—A Manual of Seafood Products, Marketing, and Utilization.* Boston, Massachusetts: Massachusetts Division of Marine Fisheries.

Gousset, J. and G. Tixerant. *Les Produits de la Peche: Poissons—Crustaces—Mollusques.* Paris: Ministère de l'Agriculture.

Grant, E.M. 1985. *Guide to Fishes.* Brisbane, Australia: Department of Harbours and Marine.

Gulland, J.A. 1970. *The Fish Resources of the Ocean.* Rome, Italy: Food and Agriculture Organisation of the United Nations.

Hart, J.L. 1988. *Pacific Fishes of Canada.* Ottawa, Canada: Department of Fisheries and Oceans.

Hedeen, Robert A. 1986. *The Oyster: The Life and Lore of the Celebrated Bivalve.* Centreville, Maryland: Tidewater Publishers.

Hines, Neal O. 1976. *Fish of Rare Breeding.* Washington, DC: Smithsonian Institution Press.

Hoese, H. Dickson and Richard H. Moore. 1977. *Fishes of the Gulf of Mexico.* College Station, Texas: Texas A & M University Press.

Howell, Thomas L. and LeGrande R. Howell. 1989. *The Controlled Purification Manual.* Boston, Massachusetts: New England Fisheries Development Association, Inc.

Howorth, Peter C. 1978. *The Abalone Book.* Happy Camp, California: Naturegraph Publishers, Inc.

Huet, Marcel. 1986. *Textbook of Fish Culture.* Farnham, Surrey, England: Fishing News Books Ltd.

Hurlburt, Sarah. 1977. *The Mussel Cookbook.* Cambridge, Massachusetts: Harvard University Press.

Jamieson, G.S. and K. Francis (eds.) 1986. Invertebrate and marine plant fishery resources of British Columbia. *Canadian Special Publication of Fisheries and Aquatic Sciences* 91:1-89.

Jensen, Chuck. 1987. *White Fish Processing Manual.* Fairbanks, Alaska: University of Alaska.

Jhaveri, Sudip et al. 1978. *Abstracts of Methods Used to Assess Fish Quality.* Narragansett, Rhode Island: University of Rhode Island.

Joseph, James, Witold Klawe and Pat Murphy. 1988. *Tuna and Billfish.* La Jolla, California: Inter-American Tropical Tuna Commission.

Krane, Willibald. 1986. *Fish: Five-language Dictionary of Fish, Crustaceans and Molluscs.* New York: Van Nostrand Reinhold.

Lamb, Andy and Phil Edgell. 1986. *Coastal Fishes of the Pacific Northwest.* Madiera Park, British Columbia: Harbour Publishing.

Lappin, Peter J. 1986. *Live Holding Systems—A Guide and Reference Manual.* Salem, Massachusetts: Sea Plantations, Inc.

Light, S.F. 1957. *Intertidal Invertebrates of the Central California Coast.* Berkeley and Los Angeles, California: University of California Press.

Manooch, Charles S. 1988. *Fisherman's Guide—Fishes of the Southeastern United States.* Raleigh, North Carolina: North Carolina State Museum of Natural History.

BIBLIOGRAPHY

Martin, Roy E. and George J. Flick (eds). 1990. *The Seafood Industry.* New York: Van Nostrand Reinhold/Osprey Books.

McClane, A. J. 1977. *The Encyclopeadia of Fish Cookery.* New York: Holt, Rinehart and Winston.

McClane, A. J. 1974. *New Standard Fishing Encyclopeadia.* New York: Holt, Rinehart and Winston.

Migdalski, Edward C. and George S. Fichter. 1983. *The Fresh and Salt Water Fishes of the World.* New York: Greenwich House.

Millar, R.H. 1961. *Scottish Oyster Investigations 1946–1958.* UK: Her Majesty's Stationery Office.

Mosimann, Anton and Holger Hofmann. 1987. *Shellfish.* New York: William Morrow & Company Inc.

Nelson, Joseph S. 1984. *Fishes of the World.* New York: John Wiley & Sons Inc.

Nettleton, Joyce A. 1985. *Seafood Nutrition: Facts, Issues and Marketing of Nutrition in Fish and Shellfish.* New York: Van Nostrand Reinhold/Osprey Books.

Nettleton, Joyce A. 1987. *Seafood and Health.* New York: Van Nostrand Reinhold/Osprey Books.

Olson, Robert E. 1987. Marine fish parasites of public health importance. In *Seafood Quality Determination,* ed. Donald E. Kramer and John Liston pp. 339–355. New York: Elsevier Science Publishing Company Inc.

Organisation for Economic Co-operation and Development. 1990. *Multilingual Dictionary of Fish and Fish Products.* UK: Fishing News Books.

Otwell, W. Steven and John A. Koburger. 1985. *Self-Regulation Guide for Calico Scallop Processing.* Tampa, Florida: Gulf and South Atlantic Fisheries Development Foundation, Inc.

Otwell, W. Steven et al. 1984. *Quality Control in Calico Scallop Production.* Tampa, Florida: Gulf and South Atlantic Fisheries Development Foundation, Inc.

Paquette, Gerald N. 1983. *Fish Quality Improvement—A Manual for*

Plant Operators. New York: Van Nostrand Reinhold/Osprey Books.

Paust, Brian and Ronald Smith. 1986. *Salmon Shark Manual*. Fairbanks, Alaska: Alaska Sea Grant College Program, University of Alaska.

Pennington, Jean A. T. and Helen Nichols Church. 1985. *Food Values of Portions Commonly Used*. New York: Harper and Row.

Price, Robert J. and Pamela Tom (eds.) 1990. *Menu and Advertising Guidelines for California Restaurants, Retailers, and Their Seafood Suppliers*. Davis, California: Food Science and Technology, University of California.
Poissons et Fruits de Mer de France. Paris, France: F.I.O.M.

Regenstein, Joe and Carrie. *Old Laws in a New Market, The Kosher Dietary Laws for Seafood Processors*. New York: New York Seagrant Institute.

Robins, C. Richard et al. 1980. *A List of Common and Scientific Names of Fishes from the United States and Canada*. Bethesda, Maryland: American Fisheries Society.

Romashko, Sandra. 1977. *The Savory Shellfish of North America*. Miami, FL: Windward Publishing Inc.

Scott, J.S. 1959. *An Introduction to the Sea Fishes of Malaya*. Kuala Lumpur, Malaya: Ministry of Agriculture, Federation of Malaya.

Scott, W.B. and M.G. Scott. 1988. *Atlantic Fishes of Canada*. Toronto, Ontario: University of Toronto Press.

Scott, W.B. and E.J. Crossman. 1985. *Freshwater Fishes of Canada*. Ottawa: Fisheries Research Board of Canada.

Seafood Product Quality Code. Tallahassee, Florida: Southeastern Fisheries Association Inc.

Sedgwick, S. Drummond. 1982. *The Salmon Handbook*. London, England: André Deutsch Ltd.

Slabyj, Bohdan M. and Gilles R. Bolduc. 1987. Applicability of commercial testing kits for microbiological quality control of seafoods. In *Seafood Quality Determination*, ed. Donald E. Kramer and John Liston, pp. 255–267. New York: Elsevier Science Publishing Company Inc.

BIBLIOGRAPHY

Turgeon, Donna D. et al. 1988. *Common and Scientific Names of Aquatic Invertebrates from the United States and Canada.* Bethesda, Maryland: American Fisheries Society.

Walford, Lionel A. 1974. *Marine Game Fishes of the Pacific coast from Alaska to the Equator.* Washington, DC: Smithsonian Institution Press.

Wheeler, Alwyne. 1975. *Fishes of the World—An Illustrated Dictionary.* New York: Macmillan Publishing Co., Inc.

White, Alan W. 1983. Red tides. *Underwater World.* Ottawa, Canada: Communications Directorate of the Department of Fisheries and Oceans.

Whitehead, P.J.P. et al. (eds.) 1984. *Fishes of the Northeastern Atlantic and the Mediterranean Volume I.* Paris, France: UNESCO.

Whitehead, P.J.P. et al. (eds.) 1986. *Fishes of the Northeastern Atlantic and the Mediterranean Volume II.* Paris, France: UNESCO.

Whitehead, P.J.P. et al. (eds.) 1986. *Fishes of the Northeastern Atlantic and the Mediterranean Volume III.* Paris, France: UNESCO.

Williams, Austin B. et al. 1989. *Decapod Crustaceans.* Bethesda, Maryland: American Fisheries Society.

Williams, Austin B. 1988. *Lobsters of the World - An Illustrated Guide.* New York: Van Nostrand Reinhold/Osprey Books.

Yonge, C. M. 1966. *Oysters.* UK: Collins Clear-Type Press.

Yoshino, Masuo. 1986. *Sushi.* Tokyo, Japan: Gakken Co., Ltd.

Yudkin, John. 1986. *The Penguin Encyclopeadia of Nutrition.* UK: Penguin Books Ltd.

Zinn, Donald J. 1976. *The Handbook for Beach Strollers from Maine to Cape Hatteras.* Chester, Connecticut: The Pequot Press.

Appendix

The following list is based upon the FDA's 1988 publication *The Fish List: FDA Guide to Acceptable Market Names for the Food Fish Sold in Interstate Commerece*. We have corrected a number of typographical errors and updated it for names approved by the FDA since 1988. We have added the scientific family names (in English and in Latin). These help you to identify where unlisted species might fit and what names are used for related fish. We have also added a code showing the location of the species: A stands for Atlantic or Gulf waters; P means Pacific; and F is for fresh water.

If you have a fish that does not appear on the list and want to establish a market name for it in the United States, follow the priorities used by the government in drawing up the *Fish List*. The highest priority rules. For example, if there is no federal law concerning the name of the species, but a name is listed by the American Fisheries Society, then the AFS name is the one you have to use. Always secure the FDA's approval before using a name. The priorities are:

1. A common or usual name established by federal law or regulation.
2. A common name selected and listed by the American Fisheries Society in its *List of Common and Scientific Names of Fishes from the United States and Canada* or in the updated 1991 version *Worldwide Listing of Fishes and Invertebrates*.
3. A common name most often cited in authoritative literature.
4. For species from foreign waters being marketed in the United States, a market name most often used in international marketing or a common name cited in the FAO's *Yearbook of Fishery Statistics* for the species, with preference given to an English name or the English translation of a foreign language name.
5. A market name most often applied historically by the industry in marketing the species.
6. If the species has not established a domestic market name or for-

APPENDIX

eign common name, a market name which describes the species as long as it does not trade on the established market name of another species.

7. The market name used for other species of the same genus, or other genera in the same scientific family.

Note that to date there is no comparable listing for shellfish.

The Fish List:
Approved Names
For Fish

The Fish List: Approved Names for Fish

Market Name	Common Name	Scientific Name	Family (Latin)	Family (English)	Location
Aholehole	Flagtail, spotted	Kuhlia marginata	Kuhliidae	Aholehole	P
Aholehole	Flagtail, rock	Kuhlia rupestris	Kuhliidae	Aholehole	P
Aholehole	Aholehole, Hawaiian	Kuhlia sandvicensis	Kuhliidae	Aholehole	P
Alewife/river herring	Alewife	Alosa pseudoharengus	Clupeidae	Herring	A-F
Alfonsin	Bream, red	Beryx decadactylus	Berycidae	Alfonsin	A
Alfonsin	Alfonsin a casta	Beryx splendens	Berycidae	Alfonsin	P
Alfonsin	Alfonsin	Trachichthodes affinis	Berycidae	Alfonsin	P
Amberjack	Amberjack, greater	Seriola dumerili	Carangidae	Jack	A
Amberjack	Amberjack, lesser	Seriola fasciata	Carangidae	Jack	A
Amberjack	Rudderfish, banded	Seriola zonata	Carangidae	Jack	A
Amberjack/yellowtail	Yellowtail	Seriola lalandei	Carangidae	Jack	P
Amberjack/yellowtail	Amberjack, king	Seriola quinqueradiata	Carangidae	Jack	P
Anchovy	Anchovy, key	Anchoa cayorum	Engraulidae	Anchovy	A
Anchovy	Anchovy, deepbody	Anchoa compressa	Engraulidae	Anchovy	P
Anchovy	Anchovy, slough	Anchoa delicatissima	Engraulidae	Anchovy	P
Anchovy	Anchovy, New Jersey	Anchoa duodecim	Engraulidae	Anchovy	A
Anchovy	Anchovy, striped	Anchoa hepsetus	Engraulidae	Anchovy	A
Anchovy	Anchovy, dusky	Anchoa lyolepis	Engraulidae	Anchovy	A
Anchovy	Anchovy, bay	Anchoa mitchilli	Engraulidae	Anchovy	A
Anchovy	Anchovy, longnose	Anchoa nasuta	Engraulidae	Anchovy	A
Anchovy	Anchovy, camiguana	Anchoviella estauqual	Engraulidae	Anchovy	A
Anchovy	Anchovy	Anchoviella letidentostole	Engraulidae	Anchovy	A
Anchovy	Anchovy, flat	Anchoviella perfasciata	Engraulidae	Anchovy	A
Anchovy	Anchoveta	Cetengraulis mysticetus	Engraulidae	Anchovy	P
Anchovy	Anchovy, southern	Engraulis australis	Engraulidae	Anchovy	P
Anchovy	Anchovy, European	Engraulis encrasicolus	Engraulidae	Anchovy	A
Anchovy	Anchovy, silver	Engraulis eurystole	Engraulidae	Anchovy	A
Anchovy	Anchovy, Japanese	Engraulis japonica	Engraulidae	Anchovy	P
Anchovy	Anchovy, northern	Engraulis mordax	Engraulidae	Anchovy	P

Market Name	Common Name	Scientific Name	Family (Latin)	Family (English)	Location
Anchovy	Anchoveta	Engraulis ringens	Engraulidae	Anchovy	P
Anchovy	Anchovy, round headed	Stolephorus buccaneeri	Engraulidae	Anchovy	P
Anchovy	Anchovy, slim	Stolephorus niarcha	Engraulidae	Anchovy	A
Anchovy	Anchovy	Stolephorus purpureus	Engraulidae	Anchovy	P
Angelfish	Angelfish, blue	Holacanthus bermudensis	Pomacanthidae	Angelfish	A
Angelfish	Angelfish, queen	Holacanthus ciliaris	Pomacanthidae	Angelfish	A
Angelfish	Rock beauty	Holacanthus tricolor	Pomacanthidae	Angelfish	A
Angelfish	Angelfish, gray	Pomacanthus arcuatus	Pomacanthidae	Angelfish	A
Angelfish	Angelfish, French	Pomacanthus paru	Pomacanthidae	Angelfish	A
Argentine	Argentine, silverside	Argentina elongata	Argentinidae	Smelt herring	A-P
Armourhead	Armourhead, pelagic	Pentaceros richardsoni	Pentacerotidae	Armorhead	P
Barracuda	Barracuda, Pacific	Sphyraena argentea	Sphyraenidae	Barracuda	P
Barracuda	Barracuda, great	Sphyraena barracuda	Sphyraenidae	Barracuda	A
Barracuda	Vicuda	Sphyraena ensis	Sphyraenidae	Barracuda	A
Barracuda	Guaguanche	Sphyraena guachancho	Sphyraenidae	Barracuda	A
Barracuda	Sennet, southern	Sphyraena picudilla	Sphyraenidae	Barracuda	A
Bass	Bass, Roanoke	Ambloplites cavifrons	Centrarchidae	Sunfish	F
Bass	Bass, Ozark	Ambloplites constellatus	Centrarchidae	Sunfish	F
Bass	Bass, rock	Ambloplites rupestris	Centrarchidae	Sunfish	F
Bass	Warmouth	Lepomis gulosus	Centrarchidae	Sunfish	F
Bass	Bass, smallmouth	Micropterus dolomieui	Centrarchidae	Sunfish	F
Bass	Bass, suwannee	Micropterus notius	Centrarchidae	Sunfish	F
Bass	Bass, spotted	Micropterus punctulatus	Centrarchidae	Sunfish	F
Bass	Bass, largemouth	Micropterus salmoides	Centrarchidae	Sunfish	F
Bass	Bass, Guadalupe	Micropterus treculi	Centrarchidae	Sunfish	F
Bass	Bass, white	Morone chrysops	Percichthyidae	Bass	F
Bass	Bass, sunshine	Morone chrysops X saxatilis	Percichthyidae	Bass	F
Bass	Bass, yellow	Morone mississippiensis	Percichthyidae	Bass	F

A = Atlantic, P = Pacific, F = fresh water

The Fish List: Approved Names for Fish

Market Name	Common Name	Scientific Name	Family (Latin)	Family (English)	Location
Bass	Bass, striped	*Morone saxatilis*	Percichthyidae	Bass	A-F-P
Bass	Bass, blackmouth	*Synagrops bellus*	Percichthyidae	Bass	A
Bass sea	Bass, European sea	*Dicentrarchus labrax*	Percichthyidae	Bass	A
Bass sea	Bass, black sea	*Epinephelus tauvina*	Serranidae	Seabass	P
Bass sea	Bass, Japan sea	*Lateolabrax japonicus*	Percichthyidae	Bass	P
Bass sea	Wreckfish	*Polyprion americanus*	Percichthyidae	Bass	A
Bass sea	Wreckfish	*Polyprion moene*	Percichthyidae	Bass	P
Bass, sea	Bass, Argentine sea	*Acanthistius brasilianus*	Serranidae	Seabass	A
Bass, sea	Sea bass, bank	*Centropristis ocyurus*	Serranidae	Seabass	A
Bass, sea	Sea bass, rock	*Centropristis philadelphica*	Serranidae	Seabass	A
Bass, sea	Sea bass, black	*Centropristis striata*	Serranidae	Seabass	A
Bass, sea	Grouper, speckled dwarf	*Epinephelus merra*	Serranidae	Seabass	P
Bass, sea	Bass, Peruvian Sea	*Paralabrax callaensis*	Serranidae	Seabass	P
Bass, sea	Bass, kelp	*Paralabrax clathratus*	Serranidae	Seabass	P
Bass, sea	Bass, spotted sand	*Paralabrax maculatofasciatus*	Serranidae	Seabass	P
Bass, sea	Bass, barred sand	*Paralabrax nebulifer*	Serranidae	Seabass	P
Bass, sea	Creole-fish	*Paranthias furcifer*	Serranidae	Seabass	A
Bigeye	Bigeye	*Priacanthus arenatus*	Priacanthidae	Bigeye	A
Bigeye	Bigeye, short	*Pristigenys alta*	Priacanthidae	Bigeye	A
Bluefish	Bluefish	*Pomatomus saltatrix*	Pomatomidae	Bluefish	A
Bluegill	Bluegill	*Lepomis macrochirus*	Centrarchidae	Sunfish	F
Bluenose	Cutlerfish, Antarctic	*Hyperoglyphe antarctica*	Stromateidae	Butterfish	P
Boarfish	Boarfish, deepbody	*Antigonia capros*	Caproidae	Boarfish	A
Boarfish	Boarfish, shortspine	*Antigonia combatia*	Caproidae	Boarfish	A
Boarfish	Boarfish, giant	*Paristiopterus labiosus*	Pentacerotidae	Armorhead	P
Bonefish	Bonefish	*Albula vulpes*	Albulidae	Bonefish	A-P
Bonito	Bonito, Pacific	*Sarda chiliensis*	Scombridae	Mackerel	P
Bonito	Bonito, striped	*Sarda orientalis* (& *S. velox*)	Scombridae	Mackerel	A-P
Bonito	Bonito, Atlantic	*Sarda sarda*	Scombridae	Mackerel	A

Market Name	Common Name	Scientific Name	Family (Latin)	Family (English)	Location
Bonnetmouth	Bonnetmouth	*Emmelichthys nitidus*	Emmelichthyidae	Bonnetmouth	P
Bonnetmouth	Bonnetmouth	*Plagiogenion rubiginosus*	Emmelichthyidae	Bonnetmouth	P
Bowfin	Bowfin	*Amia calva*	Amiidae	Bowfish	F
Bream	Tai, Taiwan	*Argyrops bleekeri*	Sparidae	Porgie	P
Bream	Bream, long-spined red	*Argyrops spinifer*	Sparidae	Porgie	A-P
Bream	Bream, threadfin	*Pentapodus macrurus*	Nemipteridae	Threadfin	P
Bream	Bream, gilt-headed	*Sparus auratus*	Sparidae	Porgie	A
Bream, threadfin	Bream, Japanese threadfin	*Nemipterus jcponicus*	Nemipteridae	Threadfin	P
Bream/bogue	Bogue	*Boops boops*	Sparidae	Porgie	A
Brotula	Brotula, bearded	*Brotula barbarta*	Ophidiidae	Cusk-eel	A
Buffalo	Buffalo, smallmouth	*Ictiobus bubalus*	Catostomidae	Sucker	F
Buffalo	Buffalo, bigmouth	*Ictiobus cyprinellus*	Catostomidae	Sucker	F
Buffalo	Buffalo, black	*Ictiobus niger*	Catostomidae	Sucker	F
Bullhead	Bullhead, yellow	*Ictalurus natalis*	Ictaluridae	Catfish	F
Bullhead	Bullhead, flat	*Ictalurus platycephalus*	Ictaluridae	Catfish	F
Bullhead/catfish	Bullhead, black	*Ictalurus melas*	Ictaluridae	Catfish	F
Bullhead/catfish	Bullhead, brown	*Ictalurus nebulosus*	Ictaluridae	Catfish	F
Bumper	Bumper, Atlantic	*Chloroscombrus chrysurus*	Carangidae	Jack	A
Bumper	Bumper, Pacific	*Chloroscombrus orqueta*	Carangidae	Jack	P
Burbot	Burbot	*Lota lota*	Gadidae	Codfish	F
Butterfish	Harvestfish	*Peprilus alepidotus*	Stromateidae	Butterfish	A
Butterfish	Butterfish, Gulf	*Peprilus burti*	Stromateidae	Butterfish	A
Butterfish	Pompano, Pacific	*Peprilus simillimus*	Stromateidae	Butterfish	P
Butterfish	Butterfish	*Peprilus triccanthus*	Stromateidae	Butterfish	A
Butterflyfish	Butterflyfish, spotfin	*Chaetodon ocellatus*	Chaetodontidae	Butterflyfish	A
Butterflyfish	Butterflyfish, banded	*Chaetodcn striatus*	Chaetodontidae	Butterflyfish	A
Cabrilla	Cabrilla, spotted	*Epinephelus analogus*	Serranidae	Seabass	P
Capelin	Capelin	*Mallotus villosus*	Osmeridae	Smelt	A-P

A = Atlantic, P = Pacific, F = fresh water

The Fish List: Approved Names for Fish

Market Name	Common Name	Scientific Name	Family (Latin)	Family (English)	Location
Cardinalfish	Cardinalfish, bigeye	Epigonus telescopus	Apogonidae	Cardinalfish	P
Carp	Carp, common	Cyprinus carpio	Cyprinidae	Carp	F
Catfish	Catfish, Brazilian	Brachyplatysoma vaillanti	Pimelodidae	Catfish	F
Catfish	Catfish, white	Ictalurus catus	Ictaluridae	Catfish	F
Catfish	Catfish, blue	Ictalurus furcatus	Ictaluridae	Catfish	F
Catfish	Catfish, yaqui	Ictalurus pricei	Ictaluridae	Catfish	F
Catfish	Catfish, channel	Ictalurus punctatus	Ictaluridae	Catfish	F
Catfish	Catfish, flatwhiskered	Pinirampus spp.	Ictaluridae	Catfish	F
Catfish	Coroata	Platynematichthys notatus	Ictaluridae	Catfish	F
Catfish	Caparari	Pseudoplatystoma tigrinum	Ictaluridae	Catfish	F
Catfish	Catfish, flathead	Pylodictis olivaris	Ictaluridae	Catfish	F
Catfishsea	Catfish, hardhead	Arius felis	Ariidae	Catfish sea	A-F
Catfish sea	Catfish, gafftopsail	Bagre marinus	Ariidae	Catfish sea	A
Catfish, ocean	Wolffish, northern	Anarhichas denticulatus	Anarhichadidae	Wolffish	A
Catfish, ocean	Wolffish, Atlantic	Anarhichas lupus	Anarhichadidae	Wolffish	A
Catfish, ocean	Wolffish, spotted	Anarhichas minor	Anarhichadidae	Wolffish	A
Catfish, ocean	Wolffish, Bering	Anarhichas orientalis	Anarhichadidae	Wolffish	P
Char	Char, Arctic	Salvelinus alpinus	Salmonidae	Trout	A-F-P
Chimaera	Chimaera, longnosed	Harriotta raleighana	Rhinochimaeridae	Ray	P
Chimaera	Ratfish	Hydrolagus novaezelandiae	Chimaeridae	Chimera	P
Chimaera	Ratfish	Hydrolagus spp.	Chimaeridae	Chimera	A-P
Chub	Kiyi	Coregonus kiyi	Salmonidae	Trout	F
Chub	Chub, yellow	Kyphosus incisor	Kyphosidae	Chub sea	A
Chub	Chub, Bermuda	Kyphosus sectatrix	Kyphosidae	Chub sea	A
Cisco	Cisco, shortnose	Coregonus reighardi	Salmonidae	Trout	F
Cisco/chub	Cisco, longjaw	Coregonus alpenae	Salmonidae	Trout	F
Cisco/chub	Cisco, shortjaw	Coregonus zenithicus	Salmonidae	Trout	F
Cisco/Tullibee	Cisco/lake herring	Coregonus artedii	Salmonidae	Trout	F
Cobia	Cobia	Rachycentron canadum	Rachycentridae	Cobia	A

Market Name	Common Name	Scientific Name	Family (Latin)	Family (English)	Location
Cod	Cod, toothed	Arctogadus borisovi	Gadidae	Codfish	A
Cod	Cod, polar	Arctogadus glacialis	Gadidae	Codfish	A
Cod	Cod, Artic	Boreogadus saida	Gadidae	Codfish	A-P
Cod	Cod, saffron	Eleginus gracilis	Gadidae	Codfish	P
Cod	Cod, Greenland	Gadus ogac	Gadidae	Codfish	A
Cod, morid	Cod, morid	Pseudophycis breviusculus	Moridae	Morid cod	P
Cod, morid	Cod, rock	Lotella rhacina	Moridae	Morid cod	P
Cod, morid	Cod, morid	Mora pacifica	Moridae	Morid cod	P
Cod/Alaska cod	Cod, Pacific	Gadus macrocephalus	Gadidae	Codfish	P
Cod/codfish	Cod, Atlantic	Gadus morhua	Gadidae	Codfish	A
Conger eel	Conger, bandtooth	Ariosoma bolearicum	Congridae	Conger	A
Conger eel	Conger, eel gray	Conger cinereus	Congridae	Conger	P
Eel, conger	Conger conger	Conger conger	Congridae	Conger	A
Eel, white	Conger marginatus	Conger marginatus	Congridae	Conger	P
Conger eel	Conger eel	Conger oceanicus	Congridae	Conger	A
Conger eel	Conger eel	Conger oligoporus	Congridae	Conger	P
Conger eel	Conger, manytooth	Conger triporiceps	Congridae	Conger	A
Conger eel	Conger eel	Congrina aequoreus	Congridae	Conger	P
Conger eel	Conger, catalina	Gnathophis catalinensis	Congridae	Conger	P
Conger eel	Conger, yellow	Hildebrandia flava	Congridae	Conger	A
Conger eel	Conger, whiptail	Hildebrandia gracilior	Congridae	Conger	A
Conger eel	Conger, margintail	Paraconger caudilimbatus	Congridae	Conger	A
Corvina	Corvina, shortfin	Cynoscion parvipinnis	Sciaenidae	Drum	P
Cowfish	Cowfish, honeycomb	Lactophrys polygonia	Ostraciidae		A
Cowfish	Cowfish, scrawled	Lactophrys quadricornis	Ostraciidae		A
Crappie	Crappie, white	Pomoxis annularis	Centrarchidae	Sunfish	F
Crappie	Crappie, black	Pomoxis nigromaculatus	Centrarchidae	Sunfish	F
Croaker	Croaker, white	Argyrosomus argentatus	Sciaenidae	Drum	A-P

A = Atlantic, P = Pacific, F = fresh water

The Fish List: Approved Names for Fish

Market Name	Common Name	Scientific Name	Family (Latin)	Family (English)	Location
Croaker	Cob	*Argyrosomus hololepidotus*	Sciaenidae	Drum	A-P
Croaker	Croaker, Japanese	*Argyrosomus japonicus*	Sciaenidae	Drum	P
Croaker	Croaker, white	*Argyrosomus macrophthalmus*	Sciaenidae	Drum	A-P
Croaker	Croaker, black	*Argyrosomus nibe*	Sciaenidae	Drum	P
Croaker	Croaker, blue	*Bairdiella batabana*	Sciaenidae	Drum	A
Croaker	Croaker, striped	*Bairdiella sanctaeluciae*	Sciaenidae	Drum	A
Croaker	Croaker, black	*Cheilotrema saturnum*	Sciaenidae	Drum	P
Croaker	Croaker, white	*Genyonemus lineatus*	Sciaenidae	Drum	P
Croaker	Jewfish, spotted	*Johnius spp.*	Sciaenidae	Drum	P
Croaker	Drummer, whitemouth	*Micropogonias furnieri*	Sciaenidae	Drum	A
Croaker	Croaker, Atlantic	*Micropogonias undulatus*	Sciaenidae	Drum	A-F
Croaker	Croaker, smalleye	*Nebris microps*	Sciaenidae	Drum	A-F
Croaker	Croaker	*Nibea spp.*	Sciaenidae	Drum	P
Croaker	Croaker, reef	*Odontoscion dentex*	Sciaenidae	Drum	A
Croaker	Croaker	*Pachypops spp.*	Sciaenidae	Drum	A
Croaker	Croaker	*Pachyurus spp.*	Sciaenidae	Drum	A
Croaker	Croaker	*Paralonchurus spp.*	Sciaenidae	Drum	A-P
Croaker	Croaker	*Plagioscion spp.*	Sciaenidae	Drum	A
Croaker	Captainfish	*Pseudotolithes spp.*	Sciaenidae	Drum	A
Croaker	Croaker	*Pterotolithus spp.*	Sciaenidae	Drum	P
Croaker	Croaker, spotfin	*Roncador stearnsi*	Sciaenidae	Drum	P
Croaker	Croaker, yellowfin	*Umbrina roncador*	Sciaenidae	Drum	P
Croaker/corvina	Corvina, gulf	*Cynoscion orthonopterus*	Sciaenidae	Drum	P
Croaker/corvina	Corvina, bigtooth	*Isopisthus parvipinnis*	Sciaenidae	Drum	A
Croaker/corvina	Corvina	*Micropogonias opercularis*	Sciaenidae	Drum	A
Croaker/shadefish	Shadefish, Atlantic	*Argyrosomus regius*	Sciaenidae	Drum	A
Croaker/yellowfish	Croaker, yellow	*Pseudosciaena spp.*	Sciaenidae	Drum	P
Cunner	Cunner	*Tautogolabrus adspersus*	Labridae	Wrasse	A
Cusk	Cusk	*Brosme brosme*	Gadidae	Codfish	A

230

Market Name	Common Name	Scientific Name	Family (Latin)	Family (English)	Location
Cusk-eel	Cusk-eel, fawn	Lepophidium cervinum	Ophidiidae	Cusk-eel	A
Cusk-eel	Cusk-eel, blackedge	Lepophidium graellsi	Ophidiidae	Cusk-eel	A
Cutlassfish	Scabbardfish, black	Aphanopus carbo	Trichiuridae	Cutlassfish	A
Cutlassfish	Scabbardfish, silver	Lepidopus caudatus	Trichiuridae	Cutlassfish	A-P
Cutlassfish	Scabbardfish	Lepidopus caudatus	Trichiuridae	Cutlassfish	P
Cutlassfish	Cutlassfish, Atlantic	Trichiurus lepturus	Trichiuridae	Cutlassfish	A
Cutlassfish	Cutlassfish, Pacific	Trichiurus nitens	Trichiuridae	Cutlassfish	P
Cutlassfish	Hairtail, smallhead	Trichiurus savaka	Trichiuridae	Cutlassfish	P
Dogfish	Catshark	Centrophorus squamosus	Squalidae	Shark	P
Dogfish	Dogfish, black	Centroscymnus fabricii	Squalidae	Shark	A
Dogfish	Dogfish	Mustelus antarcticus	Carcharhinidae	Shark	P
Dogfish	Dogfish, smooth	Mustelus canis	Carcharhinidae	Shark	A
Dogfish	Shark, dogfish	Mustelus lenticulatus	Carcharhinidae	Shark	P
Dogfish	Dogfish, lesser spotted	Scyliorhinus canicula	Scyliorhinidae	Shark	A
Dogfish	Dogfish, large spotted	Scyliorhinus stellaris	Scyliorhinidae	Shark	A
Dogfish	Dogfish, spiny/spring*	Squalus acanthias	Squalidae	Shark	A-P
Dogfish	Dogfish, northern	Squalus blainvillei	Squalidae	Shark	A-P
Dory	Dory, silver	Cyttus novae-zelandiae	Zeidae	Dory	P
Dory	Dory, buckler	Zenopsis conchifer	Zeidae	Dory	A
Dory	Dory, mirror	Zenopsis nebulosa	Zeidae	Dory	P
Dory	John Dory, European	Zeus faber	Zeidae	Dory	P
Driftfish	Driftfish, black	Hyperoglyphe bythites	Stromateidae	Butterfish	A
Driftfish	Barrelfish	Hyperoglyphe perciformis	Stromateidae	Butterfish	A
Drum	Drum, spotted	Equetus punctatus	Sciaenidae	Drum	A
Drum	Drum	Larimus spp.	Sciaenidae	Drum	A
Drum	Drum, black	Pogonias cromis	Sciaenidae	Drum	A
Drum	Drum	Stellifer spp.	Sciaenidae	Drum	A
Drum	Drum, sand	Umbrina corides	Sciaenidae	Drum	A

A = Atlantic, P = Pacific, F = fresh water
* The use of Cape shark for dogfish has been approved since the *List* was originally published.

The Fish List: Approved Names for Fish

Market Name	Common Name	Scientific Name	Family (Latin)	Family (English)	Location
Drum, freshwater	Drum, freshwater	*Aplodinotus grunniens*	Sciaenidae	Drum	F
Drum/cubbyu	Cubbyu	*Equetus umbrosus*	Sciaenidae	Drum	A
Drum/lion fish	Lion fish	*Collichthys spp.*	Sciaenidae	Drum	P
Drum/meagre	Drum/meagre	*Sciaena spp.*	Sciaenidae	Drum	A
Drum/queenfish	Queenfish	*Seriphus politus*	Sciaenidae	Drum	P
Drum/redfish	Drum, red	*Sciaenops ocellatus*	Sciaenidae	Drum	A-F
Drummer	Drummer, lowfinned	*Kyphosus vaigiensis*	Kyphosidae	Chub sea	P
Eel	Eel, European	*Anguilla anguilla*	Anguillidae	Eel	A-F
Eel	Eel, short-fin	*Anguilla australis*	Anguillidae	Eel	P-F
Eel	Eel, long-finned	*Anguilla dieffenbachii*	Anguillidae	Eel	P-F
Eel	Eel, Japanese	*Anguilla japonicus*	Anguillidae	Eel	P-F
Eel, freshwater	Eel, American	*Anguilla rostrata*	Anguillidae	Eel	A-F
Eelpout	Eelpout	*Zoarces viviparus*	Zoarcidae	Eelpout	A
Eelpout/ocean pout	Pout, ocean	*Macrozoarces americanus*	Zoarcidae	Eelpout	A
Elephant fish	Fish, elephant	*Callorhynchus millii*	Callorhynchidae	Elephant fish	P
Emperor	Emperor	*Lethrinus spp.*	Lethrinidae	Emperor	P
Fanfish	Fanfish, Pacific	*Pteraclis aesticola*	Bramidae	Pomfret	P
Filefish	Filefish, unicorn	*Aluterus monoceros*	Balistidae	Leather jacket	A
Filefish	Filefish, orange	*Aluterus schoepfi*	Balistidae	Leatherjacket	A
Flounder	Flounder, three-eye	*Ancylopsetta dilecta*	Bothidae	Flatfish	P
Flounder	Flounder	*Arnoglossus scapha*	Bothidae	Flatfish	P
Flounder	Flounder, kamchatka	*Atheresthes evermanni*	Pleuronectidae	Flatfish	P
Flounder	Flounder, peacock	*Bothus lunatus*	Bothidae	Flatfish	A
Flounder	Flounder, tropical	*Bothus mancus*	Bothidae	Flatfish	A
Flounder	Flounder, eyed	*Bothus ocellatus*	Bothidae	Flatfish	A
Flounder	Flounder, panther	*Bothus pantherinus*	Bothidae	Flatfish	A
Brill, New Zealand	Brill	*Colistium guntheri*	Pleuronectidae	Flatfish	P
Flounder	Flounder, Bering	*Hippoglossoides robustus*	Pleuronectidae	Flatfish	P

Market Name	Common Name	Scientific Name	Family (Latin)	Family (English)	Location
Flounder	Flounder, yellowtail	Limanda ferruginea	Pleuronectidae	Flatfish	A
Flounder	Flounder, Arctic	Liopsetta glacialis	Pleuronectidae	Flatfish	P
Flounder	Flounder, Gulf	Paralichthys albigutta	Bothidae	Flatfish	A
Flounder	Flounder, fourspot	Paralichthys oblongus	Bothidae	Flatfish	A
Flounder	Flounder, olive	Paralichthys olivaceus	Bothidae	Flatfish	A
Flounder	Flounder, Patagonian	Paralichthys patagonicus	Bothidae	Flatfish	A
Flounder	Flounder, broad	Paralichthys squamilentus	Bothidae	Flatfish	A
Flounder	Flounder, longjawed	Pelecanichthys crumenalis	Bothidae	Flatfish	P
Flounder	Sole, New Zealand lemon	Pelotretis flavilatus	Pleuronectidae	Flatfish	P
Flounder	Sole, New Zealand	Peltorhamphus novaezeelandiae	Pleuronectidae	Flatfish	P
Flounder	Flounder, starry	Platichthys stellatus	Pleuronectidae	Flatfish	P-F
Flounder	Flounder, Indian Ocean	Psettodes erumei	Psettodidae	Flatfish	P
Flounder	Flounder, largetoothed	Pseudorhombus arsius	Bothidae	Flatfish	P
Flounder	Flounder, small-toothed	Pseudorhombus jenynsii	Bothidae	Flatfish	A-P
Flounder	Flounder, fivespot	Pseudorhombus pentophthalmus	Bothidae	Flatfish	P
Flounder	Flounder, yellowbelly	Rhombosolea leporina	Pleuronectidae	Flatfish	P
Flounder	Flounder, sand	Rhombosolea plebeia	Pleuronectidae	Flatfish	P
Flounder	Flounder, black	Rhombosolea retiaria	Pleuronectidae	Flatfish	F
Flounder	Flounder, greenback	Rhombosolea tapirina	Pleuronectidae	Flatfish	P
Flounder	Flounder	Samariscus triocellatus	Pleuronectidae	Flatfish	P
Flounder	Windowpane	Scophthalmus aquosus	Bothidae	Flatfish	A
Flounder	Brill	Scophthalmus rhombus	Bothidae	Flatfish	A
Flounder arrowtooth	Flounder, arrowtooth	Atheresthes stomias	Pleuronectidae	Flatfish	P
Flounder/dab	Dab, common	Limanda limanda	Pleuronectidae	Flatfish	A
Flounder/dab	Dab, longhead	Limanda proboscidea	Pleuronectidae	Flatfish	A-P
Flounder/fluke	Flounder, summer	Paralichthys dentatus	Bothidae	Flatfish	A
Flounder/fluke	Flounder, southern	Paralichthys lethostigma	Bothidae	Flatfish	A-F
Flounder/fluke	Flounder, European	Platichthys flesus	Pleuronectidae	Flatfish	A

A = Atlantic, P = Pacific, F = fresh water

The Fish List: Approved Names for Fish

Market Name	Common Name	Scientific Name	Family (Latin)	Family (English)	Location
Flounder/sole	Flounder, winter/lemon sole	Pseudopleuronectes americanus	Pseuronectidae	Flatfish	A
Flounder/whiff	Scaldfish, fourspot	Lepidorhombus boscii	Bothidae	Flatfish	A
Flounder/whiff	Megrim	Lepidorhombus whiffiagonis	Bothidae	Flatfish	A
Flyingfish	Flyingfish, jenkins	Cypselurus atrisignis	Exocoetidae	Flyingfish	P
Flyingfish	Flyingfish, California	Cypselurus californicus	Exocoetidae	Flyingfish	P
Flyingfish	Flyingfish, margined	Cypselurus cyanopterus	Exocoetidae	Flyingfish	A
Flyingfish	Flyingfish, bandwing	Cypselurus exsiliens	Exocoetidae	Flyingfish	A
Flyingfish	Flyingfish, spotfin	Cypselurus furcatus	Exocoetidae	Flyingfish	A
Flyingfish	Flyingfish, blotchwing	Cypselurus hubbsi	Exocoetidae	Flyingfish	P
Flyingfish	Flyingfish, Atlantic	Cypselurus melanurus	Exocoetidae	Flyingfish	A
Flyingfish	Flyingfish, shortnosed	Cypselurus simus	Exocoetidae	Flyingfish	P
Flyingfish	Flyingfish, cuvier's	Cypselurus speculiger	Exocoetidae	Flyingfish	P
Flyingfish	Flyingfish, spotted wing	Cypselurus spilonopterus	Exocoetidae	Flyingfish	P
Flyingfish	Flyingfish, clearwing	Cypselurus comatus	Exocoetidae	Flyingfish	A
Flyingfish	Flyingfish, ocean two-wing	Exocoetus obtusirostris	Exocoetidae	Flyingfish	A
Flyingfish	Flyingfish, trop. two-wing	Exocoetus volitans	Exocoetidae	Flyingfish	P
Flyingfish	Flyingfish, sharpchin	Fodiator acutus	Exocoetidae	Flyingfish	A
Flyingfish	Flyingfish, fourwing	Hirundichthys affinis	Exocoetidae	Flyingfish	A-P
Flyingfish	Flyingfish, blackwing	Hirundichthys rondeleti	Exocoetidae	Flyingfish	A
Flyingfish	Flyingfish, smallwing	Oxyporhamphus micropterus	Exocoetidae	Flyingfish	A
Flyingfish	Flyingfish, sailfin	Parexocoetus brachypterus	Exocoetidae	Flyingfish	A
Flyingfish	Flyingfish, bluntnose	Prognichthys gibbifrons	Exocoetidae	Flyingfish	A
Gar	Gar, longnose	Lepisosteus osseus	Lepisosteidae	Gar	F
Gar	Gar, alligator	Lepisosteus spatula	Lepisosteidae	Gar	F
Gemfish	Domine	Epinnula magistralis	Gempylidae	Snake mackerel	A-P
Gemfish	Escolar	Lepidocybium flavobrunneum	Gempylidae	Snake mackerel	A-P
Gemfish	Gemfish, black	Nesiarchus nasutus	Gempylidae	Snake mackerel	A-P
Gemfish/barracouta	Gemfish, silver	Rexea solandri	Gempylidae	Snake mackerel	P
Gemfish/barracouta	Mackerel, snake	Thyrsites atun	Gempylidae	Snake mackerel	P

Market Name	Common Name	Scientific Name	Family (Latin)	Family (English)	Location
Gemfish/caballa	Mackerel, white snake	Thyrsites lepidopoides	Gempylidae	Snake mackerel	A
Goatfish	Goatfish	Mulloidichthys spp.	Mullidae	Goatfish	A
Goatfish	Goatfish, red	Mullus auratus	Mullidae	Goatfish	A
Goatfish	Goatfish	Parupeneus spp.	Mullidae	Goatfish	P
Goatfish	Goatfish, spotted	Pseudupeneus spp.	Mullidae	Goatfish	A
Goatfish	Goatfish, bluespotted	Upeneichthys lineatus	Mullidae	Goatfish	P
Goatfish	Goatfish	Upeneus spp.	Mullidae	Goatfish	A
Goldeye	Goldeye	Hiodon alosoides	Hiodontidae	Mooneye	F
Goldfish	Goldfish	Carassius auratus	Cyprinidae	Carp	F
Grayling	Grayling, Arctic	Thymallus arcticus	Salmonidae	Trout	F
Greenbone	Greenbone	Coridodax pullus	Odacidae	Greenbone	P
Greenling	Greenling, kelp	Hexagrammos decagrammus	Hexagrammidae	Greenling	P
Greenling	Greenling, rock	Hexagrammos lagocephalus	Hexagrammidae	Greenling	P
Greenling	Greenling, masked	Hexagrammos octogrammus	Hexagrammidae	Greenling	P
Greenling	Greenling, whitespotted	Hexagrammos stelleri	Hexagrammidae	Greenling	P
Grenadier	Rattail, giant	Coryphaenoides pectoralis	Macrouridae	Grenadier	P
Grenadier	Grenadier, rock	Coryphaenoides rupestris	Macrouridae	Grenadier	A
Grenadier	Whiptail, deepsea	Lepidorhynchus denticulatus	Macrouridae	Grenadier	P
Grenadier	Grenadier	Macrourus spp.	Macrouridae	Grenadier	P
Grenadier	Marlin-spike	Nezumia bairdi	Macrouridae	Grenadier	A
Grenadier	Grenadier, roughnose	Trachyrhynchus murrayi	Macrouridae	Grenadier	A
Grouper	Grouper	Caprodon schlegelii	Serranidae	Seabass	P
Grouper	Coney, gulf	Cephalopholis acanthistius	Serranidae	Seabass	P
Grouper	Grouper, purplespotted	Cephalopholis argus	Serranidae	Seabass	P
Grouper	Grouper, spotted	Cephalopholis taeniops	Serranidae	Seabass	A
Grouper	Grouper, chevron tailed	Cephalopholis urodelus	Serranidae	Seabass	P
Grouper	Perch, sand	Diplectrum formosum	Serranidae	Seabass	A
Grouper	Grouper, white	Epinephelus aeneus	Serranidae	Seabass	A

A = Atlantic, P = Pacific, F = fresh water

The Fish List: Approved Names for Fish

Market Name	Common Name	Scientific Name	Family (Latin)	Family (English)	Location
Grouper	Rockcod, yellowspotted	*Epinephelus areolatus*	Serranidae	Seabass	A-P
Grouper	Rockcod, brownspotted	*Epinephelus chlorostigma*	Serranidae	Seabass	A-P
Grouper	Graysby	*Epinephelus cruentatus*	Serranidae	Seabass	A
Grouper	Grouper, blacktip	*Epinephelus fasciatus*	Serranidae	Seabass	P
Grouper	Grouper, yellowedge	*Epinephelus flavolimbatus*	Serranidae	Seabass	A
Grouper	Coney	*Epinephelus fulva*	Serranidae	Seabass	A
Grouper	Grouper, mottled	*Epinephelus fuscoguttatus*	Serranidae	Seabass	P
Grouper	Grouper, dusky	*Epinephelus guaza*	Serranidae	Seabass	A
Grouper	Grouper, marbled	*Epinephelus inermis*	Serranidae	Seabass	A
Grouper	Grouper, red	*Epinephelus morio*	Serranidae	Seabass	A
Grouper	Grouper, misty	*Epinephelus mystacinus*	Serranidae	Seabass	A
Grouper	Grouper, warsaw	*Epinephelus nigritus*	Serranidae	Seabass	A
Grouper	Grouper, snowy	*Epinephelus niveatus*	Serranidae	Seabass	A-P
Grouper	Grouper, nassau	*Epinephelus striatus*	Serranidae	Seabass	A
Grouper	Grouper, black	*Mycteroperca bonaci*	Serranidae	Seabass	A
Grouper	Grouper, yellowmouth	*Mycteroperca interstitialis*	Serranidae	Seabass	A
Grouper	Grouper, gulf	*Mycteroperca jordani*	Serranidae	Seabass	P
Grouper	Grouper, comb	*Mycteroperca rubra*	Serranidae	Seabass	A-P
Grouper	Grouper, tiger	*Mycteroperca tigris*	Serranidae	Seabass	A
Grouper	Grouper, yellowfin	*Mycteroperca venenosa*	Serranidae	Seabass	A
Grouper	Grouper, broomtail	*Mycteroperca xenarcha*	Serranidae	Seabass	P
Grouper/gag	Gag	*Mycteroperca microlepis*	Serranidae	Seabass	A
Grouper/hind	Hind, red	*Epinephelus guttatus*	Serranidae	Seabass	A
Grouper/jewfish	Jewfish	*Epinephelus itajara*	Serranidae	Seabass	A
Grunion	Grunion, California	*Leuresthes tenuis*	Atherinidae	Silverside	P
Grunt	Grunt, burrito	*Anisotremus interruptus*	Haemulidae	Grunt	A
Grunt	Grunt, barred	*Conodon nobilis*	Haemulidae	Grunt	A-P
Grunt	Grunt, black	*Haemulon bonariense*	Haemulidae	Grunt	A
Grunt	Grunt, caesar	*Haemulon carbonarium*	Haemulidae	Grunt	A

236

Market Name	Common Name	Scientific Name	Family (Latin)	Family (English)	Location
Grunt	Grunt, smallmouth	*Haemulon chrysargyreum*	Haemulidae	Grunt	A
Grunt	Grunt, French	*Haemulon flavolineatum*	Haemulidae	Grunt	A
Grunt	Grunt, Spanish	*Haemulon macrostomum*	Haemulidae	Grunt	A
Grunt	Sailors choice	*Haemulon parrai*	Haemulidae	Grunt	A
Grunt	Grunt, white	*Haemulon plumieri*	Haemulidae	Grunt	A
Grunt	Grunt, bluestriped	*Haemulon sciurus*	Haemulidae	Grunt	A
Grunt	Prieta mojarra	*Haemulon scudderii*	Haemulidae	Grunt	P
Grunt	Grunt, Latin	*Haemulon steindachneri*	Haemulidae	Grunt	A-P
Grunt	Grunt, striped	*Haemulon striatum*	Haemulidae	Grunt	A
Grunt	Pigfish	*Orthopristis chrysoptera*	Haemulidae	Grunt	A-F
Grunt	Grunt, burro	*Pomadasys crocro*	Haemulidae	Grunt	A
Grunt/catalina	Catalina	*Anisotremus taeniatus*	Haemulidae	Grunt	A
Grunt/cottonwick	Cottonwick	*Haemulon melanurum*	Haemulidae	Grunt	A
Grunt/margate	Margate, black	*Anisotremus surinamensis*	Haemulidae	Grunt	A
Grunt/margate	Margate	*Haemulon album*	Haemulidae	Grunt	A
Grunt/porkfish	Porkfish	*Anisotremus virginicus*	Haemulidae	Grunt	A
Grunt/sargo	Sargo	*Anisotremus davidsoni*	Haemulidae	Grunt	P
Grunt/sweetlips	Sweetlips, harlequin	*Plectorhynchus chaetodonoides*	Haemulidae	Grunt	P
Grunt/sweetlips	Sweetlips, yellowbanded	*Plectorhynchus lineatus*	Haemulidae	Grunt	P
Grunt/sweetlips	Sweetlips, brown	*Plectorhynchus nigrus*	Haemulidae	Grunt	P
Grunt/sweetlips	Sweetlips, Oriental	*Plectorhynchus orientalis*	Haemulidae	Grunt	P
Grunt/sweetlips	Sweetlips, painted	*Plectorhynchus pictus*	Haemulidae	Grunt	P
Grunt/tomtate	Tomtate	*Haemulon aurolineatum*	Haemulidae	Grunt	A
Grunt/tomtate	Tomtate	*Haemulon rimator*	Haemulidae	Grunt	A
Guitarfish	Thornback	*Platyrhinoidis triseriata*	Rhinobatidae	Guitarfish	P
Guitarfish	Guitarfish, Atlantic	*Rhinobatos lentiginosus*	Rhinobatidae	Guitarfish	A
Guitarfish	Guitarfish, shovelnose	*Rhinobatos productus*	Rhinobatidae	Guitarfish	P
Guitarfish	Guitarfish, banded	*Zapteryx exasperata*	Rhinobatidae	Guitarfish	P

A = Atlantic, P = Pacific, F = fresh water

The Fish List: Approved Names for Fish

Market Name	Common Name	Scientific Name	Family (Latin)	Family (English)	Location
Haddock	Haddock	Melanogrammus aeglefinus	Gadidae	Codfish	A
Hake	Hake, longfin	Phycis chesteri	Gadidae	Codfish	A
Hake	Hake, Brazilian	Urophycis brasiliensis	Gadidae	Codfish	A
Hake	Hake, red	Urophycis chuss	Gadidae	Codfish	A
Hake	Hake, Gulf	Urophycis cirrata	Gadidae	Codfish	A
Hake	Hake, Carolina	Urophycis earlli	Gadidae	Codfish	A
Hake	Hake, southern	Urophycis floridana	Gadidae	Codfish	A
Hake	Hake, spotted	Urophycis regia	Gadidae	Codfish	A
Hake	Hake, white	Urophycis tenuis	Gadidae	Codfish	A
Halfmoon	Halfmoon	Medialuna californiensis	Kyphosidae	Chub sea	P
Halibut	Halibut, Atlantic	Hippoglossus hippoglossus	Pleuronectidae	Flatfish	A
Halibut	Halibut, Pacific	Hippoglossus stenolepis	Pleuronectidae	Flatfish	P
Halibut/California halibut	Halibut, California	Paralichthys californicus	Bothidae	Flatfish	P
Hamlet	Hamlet, mutton	Epinephelus afer	Serranidae	Seabass	A
Herring	Herring, round	Etrumeus teres	Clupeidae	Herring	A-P
Herring	Ilisha, African	Ilisha africans	Clupeidae	Herring	A
Herring	Ilisha, pugnose	Ilisha elongata	Clupeidae	Herring	P
Herring	Ilisha, Pacific	Ilisha furthi	Clupeidae	Herring	P
Herring	Ilisha, bigeye	Ilisha megaloptera	Clupeidae	Herring	P
Herring	Ilisha, Indian	Ilisha melastoma	Clupeidae	Herring	P
Herring	Ilisha, javan	Ilisha pristgastroides	Clupeidae	Herring	P
Herring	Tardoore	Opisthopterus tardoore	Clupeidae	Herring	P
Herring	Pellona, Indian	Pellona ditchela	Clupeidae	Herring	P
Herring, thread	Thread herring, Atlantic	Opisthonema oglinum	Clupeidae	Herring	A-F
Herring/river herring	Herring, blueback	Alosa aestivalis	Clupeidae	Herring	A-F
Herring/river herring	Herring, skipjack	Alosa chrysochloris	Clupeidae	Herring	A-P
Herring/sea herring/sild	Herring, Atlantic	Clupea harengus harengus	Clupeidae	Herring	A
Herring/sea herring/sild	Herring, Pacific	Clupea harengus pallasi	Clupeidae	Herring	P
Hind	Hind, rock	Epinephelus adscensionis	Serranidae	Seabass	A

238

Market Name	Common Name	Scientific Name	Family (Latin)	Family (English)	Location
Hind	Hind, speckled	Epinephelus drummondhayi	Serranidae	Seabass	A
Hogfish	Hogfish, spotfin	Bodianus pulchellus	Labridae	Wrasse	A
Hogfish	Hogfish, Spanish	Bodianus rufus	Labridae	Wrasse	A
Hogfish	Hogfish	Lachnolaimus maximus	Labridae	Wrasse	A
Houndfish	Houndfish	Tylosurus crocodilus	Belonidae	Needlefish	A
Icefish	Icefish, mackerel	Champsocephalus gunnari	Channichthyidae	Morwong	A
Icefish	Icefish	Notothenia moariensis	Nototheniidae	Cod icefish	P
Icefish	Icefish, Japanese	Salangichthys spp.	Salangidae	Icefish	P
Jack	Jack, yellow	Caranx bartholomaei	Carangidae	Jack	A
Jack	Jack, green	Caranx caballus	Carangidae	Jack	P
Jack	Jack, whitemouth	Caranx helvolus	Carangidae	Jack	P
Jack	Jack, crevalle	Caranx hippos	Carangidae	Jack	A
Jack	Jack, horse-eye	Caranx latus	Carangidae	Jack	A
Jack	Jack, black	Caranx lugubris	Carangidae	Jack	A
Jack	Jack, bar	Caranx ruber	Carangidae	Jack	A
Jack	Jack, mazatlan/striped	Caranx vinctus	Carangidae	Jack	P
Jack	Leather jacket	Oligoplites saurus	Carangidae	Jack	A
Jack	Moonfish, Pacific	Selene peruviana	Carangidae	Jack	P
Jack	Moonfish, Atlantic	Selene setapinnis	Carangidae	Jack	A
Jack	Lookdown	Selene vomer	Carangidae	Jack	A
Jack	Jack, almaco	Seriola rivoliana	Carangidae	Jack	A
Jack	Jack, cottonmouth	Uraspis secunda	Carangidae	Jack	A-P
Jack mackerel	Mackerel, jack	Trachurus declivis	Carangidae	Jack	A
Jack mackerel	Scad, Japanese	Trachurus japonicus	Carangidae	Jack	P
Jack mackerel	Scad, rough	Trachurus iathami	Carangidae	Jack	A
Jack mackerel	Scad, inca	Trachurus murphyi	Carangidae	Jack	P
Jack mackerel	Mackerel, jack	Trachurus symmetricus	Carangidae	Jack	P
Jack/blue runner	Runner, blue	Caranx crysos	Carangidae	Jack	A

A = Atlantic, P = Pacific, F = fresh water

239

The Fish List: Approved Names for Fish

Market Name	Common Name	Scientific Name	Family (Latin)	Family (English)	Location
Jack/rainbow runner	Runner, rainbow	*Elegatis bipinnulata*	Carangidae	Jack	A-P
Jack/rooster fish	Roosterfish	*Nematistius pectoralis*	Carangidae	Jack	P
Jack/trevally	Trevally, white	*Caranx georgianus*	Carangidae	Jack	P
Jack/trevally	Trevally, giant	*Caranx ignobilis*	Carangidae	Jack	P
Jack/trevally	Jack, spotted	*Caranx melampygus*	Carangidae	Jack	P
Jackfish/crevalle	Jack, threadfin	*Alectis indicus*	Carangidae	Jack	P
Jobfish	Jobfish, smalltooth	*Aphareus furcatus*	Lutjanidae	Snapper	P
Jobfish	Jobfish, rusty	*Aphareus rutilans*	Lutjanidae	Snapper	P
Jobfish	Jobfish, green	*Aprion virescens*	Lutjanidae	Snapper	P
Jobfish	Snapper/jobfish goldeneye	*Pristipomoides flavipinnis*	Lutjanidae	Snapper	P
Jobfish	Jobfish, lavender	*Pristipomoides sieboldii*	Lutjanidae	Snapper	P
Kahawai	Kahawai	*Arripis trutta*	Arripididae	Kahawai	P
Kelpfish	Kelpfish	*Chironemus marmoratus*	Chironemidae	Kelpfish	P
Killifish	Killifish, marsh	*Fundulus confluentus*	Cyprinodontidae	Killifish	A-F
Killifish	Mummichog	*Fundulus heteroclitus*	Cyprinodontidae	Killifish	A-F
Killifish	Killifish, striped	*Fundulus majalis*	Cyprinodontidae	Killifish	A
Kingfish	Kingfish, southern	*Menticirrhus americanus*	Sciaenidae	Drum	A
Kingfish	Kingfish, Gulf	*Menticirrhus littoralis*	Sciaenidae	Drum	A
Kingfish	Kingfish, northern	*Menticirrhus saxatilis*	Sciaenidae	Drum	A
Kingfish/corbina	Corbina, California	*Menticirrhus undulatus*	Sciaenidae	Drum	P
Kingklip	Kingklip, golden	*Genypterus blacodes*	Ophidiidae	Cusk-eel	P
Kingklip	Kingklip, South African	*Genypterus capensis*	Ophidiidae	Cusk-eel	A
Kingklip	Kingklip, red	*Genypterus chilensis*	Ophidiidae	Cusk-eel	A
Kingklip	Kingklip, black	*Genypterus maculatus*	Ophidiidae	Cusk-eel	P
Ladyfish	Machete	*Elops affinis*	Elopidae	Tarpon	P-F
Ladyfish	Ladyfish	*Elops hawaiensis*	Elopidae	Tarpon	P-F
Ladyfish	Ladyfish	*Elops saurus*	Elopidae	Tarpon	A-F
Ling	Ling, blue	*Molva dypterygia*	Gadidae	Codfish	A
Ling	Ling	*Molva molva*	Gadidae	Codfish	A

Market Name	Common Name	Scientific Name	Family (Latin)	Family (English)	Location
Ling Mediterranean	Ling, Spanish	Molva macrophthalus	Gadidae	Codfish	A
Lingcod	Lingcod	Ophiodon elongatus	Hexagrammidae	Greenling	P
Lizardfish	Lizardfish, inshore	Synodus foetens	Synodontidae	Lizardfish	A
Lizardfish	Sand diver	Synodus intermedius	Synodontidae	Lizardfish	A
Louvar	Louvar	Luvarus imperialis	Luvaridae	Louvar	A-P
Lumpfish	Lumpfish	Cyclopterus lumpus	Cyclopteridae	Lumpfish	A
Mackerel	Mackerel, frigate	Auxis thazard	Scombridae	Mackerel	A-P
Mackerel	Mackerel, double-lined	Grammatorcynus bicarinatus	Scombridae	Mackerel	P
Mackerel	Mackerel, Indian	Rastrelliger kanagurta	Scombridae	Mackerel	P
Mackerel	Mackerel, chub	Scomber japonicus	Scombridae	Mackerel	A-P
Mackerel	Mackerel, Atlantic	Scomber scombrus	Scombridae	Mackerel	A
Mackerel	Mackerel, narrow-barred	Scomberomorus commerson	Scombridae	Mackerel	P
Mackerel	Sierra, gulf	Scomberomorus concolor	Scombridae	Mackerel	P
Mackerel	Mackerel, Spanish	Scomberomorus maculatus	Scombridae	Mackerel	A
Mackerel	Cero	Scomberomorus regalis	Scombridae	Mackerel	A
Mackerel	Sierra, Pacific	Scomberomorus sierra	Scombridae	Mackerel	P
Mackerel	Mackerel, Spanish	Scomberomorus tritor	Scombridae	Mackerel	A
Mackerel chub	Mackerel, spotted chub	Scomber australasicus	Scombridae	Mackerel	P
Mackerel king	Mackerel, king	Scomberomorus cavalla	Scombridae	Mackerel	A
Mackerel, atka	Mackerel, atka	Pleurogrammus monopterygius	Hexagrammidae	Greenling	P
Mackerel, snake	Oilfish	Ruvettus pretiosus	Gempylidae	Snake mackerel	A-P
Mackerel/scad	Mackerel, jack	Trachurus novaezelandiae	Carangidae	Jack	P
Mahi-mahi	Dolphin, pompano	Coryphaena equisetis	Coryphaenidae	Dolphin	A
Mahi-mahi	Dolphin	Coryphaena hippurus	Coryphaenidae	Dolphin	A-P
Manta	Manta, Atlantic	Manta birostris	Mobulidae	Manta	A
Manta	Manta, Pacific	Manta hamiltoni	Mobulidae	Manta	P
Manta	Ray, devil	Mobula hypostoma	Mobulidae	Manta	A
Manta	Mobula, spinetail	Mobula japanica	Mobulidae	Manta	P

A = Atlantic, P = Pacific, F = fresh water

The Fish List: Approved Names for Fish

Market Name	Common Name	Scientific Name	Family (Latin)	Family (English)	Location
Manta	Mobula, smoothtail	*Mobula lucasana*	Mobulidae	Manta	P
Marlin	Marlin, black	*Makaira indica*	Istiophoridae	Billfish	P
Marlin	Marlin, blue	*Makaira nigricans*	Istiophoridae	Billfish	A-P
Marlin	Marlin, white	*Tetrapturus albidus*	Istiophoridae	Billfish	A
Marlin	Marlin, striped	*Tetrapturus audax*	Istiophoridae	Billfish	P
Menhaden	Menhaden, finescale	*Brevoortia gunteri*	Clupeidae	Herring	A
Menhaden	Menhaden, Gulf	*Brevoortia patronus*	Clupeidae	Herring	A
Menhaden	Menhaden, yellowfin	*Brevoortia smithi*	Clupeidae	Herring	A
Menhaden	Menhaden, Atlantic	*Brevoortia tyrannus*	Clupeidae	Herring	A
Milkfish	Milkfish	*Chanos chanos*	Chanidae	Milkfish	P
Mojarra	Pompano, Irish	*Diapterus auratus*	Gerreidae	Mojarra	A
Mojarra	Mojarra, striped	*Diapterus plumieri*	Gerreidae	Mojarra	A-F
Monkfish	Goosefish	*Lophius americanus*	Lophiidae	Goosefish	A
Monkfish	Monkfish	*Lophius piscatorius*	Lophiidae	Goosefish	P
Mooneye	Mooneye	*Hiodon tergisus*	Hiodontidae	Mooneye	F
Moray	Moray, recticulate	*Muraena retifera*	Muraenidae	Moray	A
Morwong	Marblefish	*Aplodactylus meandratus*	Cheilodactylida	Morwong	P
Morwong	Morwong, magpie	*Cheilodactylus gibbosus*	Cheilodactylida	Morwong	P
Morwong	Tarakihi	*Cheilodactylus macropterus*	Cheilodactylida	Morwong	P
Morwong	Morwong, brownband	*Cheilodactylus spectabilis*	Cheilodactylida	Morwong	P
Morwong	Morwong, blue	*Nemadactylus carponotatus*	Cheilodactylida	Morwong	P
Morwong	Morwong	*Nemadactylus douglasi*	Cheilodactylida	Morwong	P
Mullet	Mullet, mountain	*Agonostomus monticola*	Mugilidae	Mullet	A-F
Mullet	Mullet, yelloweye	*Aldrichetta forsteri*	Mugilidae	Mullet	P
Mullet	Mullet, wartynosed	*Crenimugil crenilabis*	Mugilidae	Mullet	A
Mullet	Mullet, striped	*Mugil cephalus*	Mugilidae	Mullet	A-P-F
Mullet	Mullet, white	*Mugil curema*	Mugilidae	Mullet	A-F
Mullet	Mullet, redeye	*Mugil gaimardianus*	Mugilidae	Mullet	A
Mullet	Liza	*Mugil liza*	Mugilidae	Mullet	A

Market Name	Common Name	Scientific Name	Family (Latin)	Family (English)	Location
Mullet	Mullet, fantail	Mugil trichodon	Mugilidae	Mullet	A
Mullet	Mullet, red	Mullus brabatus	Mullidae	Goatfish	A
Mullet	Mullet, red	Mullus surmuletus	Mullidae	Goatfish	A
Mullet	Mullet	Neomyxus chaptalii	Mugilidae	Mullet	P
Muskellunge	Muskellunge	Esox masquinongy	Esocidae	Pike	F
Nile perch	Perch, Nile	Lates nilotica	Centropomidae	Snook	F
Opah	Opah	Lampris guttatus	Lampridae	Opah	A-P
Opaleye	Opaleye	Girella nigricans	Kyphosidae	Chub sea	P
Oreo dory	Oreo, dory black	Allocyttus spp	Oreosomatidae	Oreo dory	P
Oreo dory	Oreo, dory smooth	Pseudocyttus maculatus	Oreosomatidae	Oreo dory	P
Oscar	Oscar	Astronotus ocellatus	Cichlidae	Cichlid	F
Paddlefish	Paddlefish	Polyodon spathula	Polyodontidae	Paddlefish	F
Pargo	Pargo, striped	Hoplopagrus guntheri	Lutjanidae	Snapper A	
Parrotfish	Parrotfish, midnight	Scarus coelestinus	Scaridae	Parrotfish	A
Parrotfish	Parrotfish, stoplight	Sparisoma viride	Scaridae	Parrotfish	A
Perch	Perch, silver	Bairdiella chrysoura	Sciaenidae	Drum	A-F
Perch	Perch, zebra	Hermosilla azurea	Kyphosidae	Chub sea	P
Perch yellow/lake perch	Perch, yellow	Perca flavescens	Percidae	Perch	F
Perch, ocean	Ocean perch, Pacific	Sebastes alutus	Scorpaenidae	Rockfish	P
Perch, ocean	Redfish, Labrador	Sebastes fasciatus	Scorpaenidae	Rockfish	A
Perch, ocean	Redfish/ocean perch	Sebastes marinus	Scorpaenidae	Rockfish	A
Perch, ocean	Deepwater, redfish	Sebastes mentella	Scorpaenidae	Rockfish	A
Perch, ocean	Redfish, Norway	Sebastes viviparus	Scorpaenidae	Rockfish	A
Perch, pile	Perch, pile	Rhacochilus vacca	Embiotocidae	Surfperch	P
Perch, white	Perch, white	Morone americana	Percichthyidae	Bass	A-F
Pickerel	Pickerel, redfin	Esox americanus americanus	Esocidae	Pike F	
Pickerel	Pickerel, grass	Esox americanus vermiculatus	Esocidae	Pike	F
Pickerel	Pickerel, chain	Esox niger	Esocidae	Pike	F

A = Atlantic, P = Pacific, F = fresh water

The Fish List: Approved Names for Fish

Market Name	Common Name	Scientific Name	Family (Latin)	Family (English)	Location
Pike	Pike, northern	*Esox lucius*	Esocidae	Pike	F
Pilchard/sardine	Pilchard, European	*Sardina pilchardus*	Clupeidae	Herring	A
Pilchard/sardine	Sardine, Australian	*Sardinops neopilchardus*	Clupeidae	Herring	P
Pilchard/sardine	Pilchard, Japanese	*Sardinops sagax melanosticta*	Clupeidae	Herring	P
Pilchard/sardine	Pilchard, South African	*Sardinops sagax ocellata*	Clupeidae	Herring	A-P
Pipefish	Pipefish, northern	*Syngnathus fuscus*	Syngnathidae	Pipefish	A
Plaice	Plaice, Alaska	*Pleuronectes quadrituberculatus*	Pleuronectidae	Flatfish	P
Plaice	Plaice, European	*Pleuronectes platessa*	Pleuronectidae	Flatfish	A
Plaice/dab	Plaice, American	*Hippoglossoides platessoides*	Pleuronectidae	Flatfish	A
Pollock	Pollock	*Pollachius pollachius*	Gadidae	Codfish	A
Pollock	Pollock	*Pollachius virens*	Gadidae	Codfish	A
Pollock/Alaska pollock	Pollock, walleye	*Theragra chalcogramma*	Gadidae	Codfish	P
Pomfret	Pomfret, Atlantic	*Brama brama*	Bramidae	Pomfret	A
Pomfret	Pomfret, Pacific	*Brama japonica*	Bramidae	Pomfret	P
Pomfret	Pomfret	*Taractes rubescens*	Bramidae	Pomfret	P
Pompano	Pompano, African	*Alectis ciliaris*	Carangidae	Jack	A
Pompano	Pompano, tropical	*Trachinotus blochi*	Carangidae	Jack	P
Pompano	Pompano, Florida	*Trachinotus carolinus*	Carangidae	Jack	A
Pompano	Pompano, paloma	*Trachinotus paitensis*	Carangidae	Jack	P
Pompano/palometa	Palometa	*Trachinotus goodei*	Carangidae	Jack	A
Pompano/permit	Permit	*Trachinotus falcatus*	Carangidae	Jack	A
Pompano/permit	Permit, palometta	*Trachinotus kennedyi*	Carangidae	Jack	P
Pompano/pompanito	Pompano, gafftopsail	*Trachinotus rhodopus*	Carangidae	Jack	P
Porgy	Porgy	*Calamus spp.*	Sparidae	Porgie	A-P
Porgy	Porgy, red Hawaiian	*Chrysophrys auratus*	Sparidae	Porgie	P
Porgy	Dentex	*Dentex gibbosus*	Sparidae	Porgie	A
Porgy	Porgy	*Diplodus spp.*	Sparidae	Porgie	A
Porgy	Pinfish	*Lagodon rhomboides*	Sparidae	Porgie	A-F
Porgy	Porgy	*Pagrus pagrus*	Sparidae	Porgie	A-P

Market Name	Common Name	Scientific Name	Family (Latin)	Family (English)	Location
Porgy	Porgy, longspine	Stenotomus caprinus	Sparidae	Porgie	A
Porgy/scup	Scup	Stenotomus chrysops	Sparidae	Porgie	A
Puffer	Puffer, smooth	Lagocephalus laevigatus	Tetraodontidae	Puffer	A
Puffer	Puffer, longnose	Sphoeroides lobatus	Tetraodontidae	Puffer	P
Puffer	Puffer, northern	Sphoeroides maculatus	Tetraodontidae	Puffer	A
Puffer	Puffer, southern	Sphoeroides nephelus	Tetraodontidae	Puffer	A
Puffer	Puffer, blunthead	Sphoeroides pachygaster	Tetraodontidae	Puffer	A
Puffer	Puffer, least	Sphoeroides parvus	Tetraodontidae	Puffer	A
Puffer	Puffer, bandtail	Sphoeroides spengleri	Tetraodontidae	Puffer	A
Puffer	Puffer, checkerd	Sphoeroides testudineus	Tetraodontidae	Puffer	A
Puffer, bullseye	Puffer, bullseye	Sphoeroides annulatus	Tetraodontidae	Puffer	P
Puffer, marbled	Puffer, marbled	Sphoeroides dorsalis	Tetraodontidae	Puffer	P
Puffer, oceanic	Puffer, oceanic	Lagocephalus lagocephalus	Tetraodontidae	Puffer	A-P
Racehorse	Pigfish, southern	Congiopodus leucopaecilus	Congiopodidae	Racehorse	P
Ray, bat	Ray, bat	Myliobatis californica	Myliobatidae	Ray eagle	P
Ray, bullnose	Ray, bullnose	Myliobatis freminvillei	Myliobatidae	Ray eagle	A
Ray, cownose	Ray, cownose	Rhinoptera bonasus	Myliobatidae	Ray eagle	A
Ray, eagle	Ray, spotted eagle	Aetobatus narinari	Myliobatidae	Ray eagel	A
Ray, eagle	Ray, southern eagle	Myliobatis goodei	Myliobatidae	Ray eagle	A
Ray, electric	Ray, lesser electric	Narcine brasiliensis	Torpedinidae	Ray electric	A
Ray, electric	Ray, Pacific electric	Torpedo californica	Torpedinidae	Ray electric	P
Rockfish	Rockfish	Helicolenus papillosus	Scorpaenidae	Rockfish	P
Rockfish	Rockfish, red	Scorpaena cardinalis	Scorpaenidae	Rockfish	P
Rockfish	Rockfish, rougheye	Sebastes aleutianus	Scorpaenidae	Rockfish	P
Rockfish	Rockfish, kelp	Sebastes atrovirens	Scorpaenidae	Rockfish	P
Rockfish	Rockfish, brown	Sebastes auriculatus	Scorpaenidae	Rockfish	P
Rockfish	Rockfish, aurora	Sebastes aurora	Scorpaenidae	Rockfish	P
Rockfish	Rockfish, redbanded	Sebastes babcocki	Scorpaenidae	Rockfish	P

A = Atlantic, P = Pacific, F = fresh water

The Fish List: Approved Names for Fish

Market Name	Common Name	Scientific Name	Family (Latin)	Family (English)	Location
Rockfish	Rockfish, shortraker	*Sebastes borealis*	Scorpaenidae	Rockfish	P
Rockfish	Rockfish, silvergray	*Sebastes brevispinis*	Scorpaenidae	Rockfish	P
Rockfish	Rockfish, gopher	*Sebastes carnatus*	Scorpaenidae	Rockfish	P
Rockfish	Rockfish, copper	*Sebastes caurinus*	Scorpaenidae	Rockfish	P
Rockfish	Rockfish, greenspotted	*Sebastes chlorostictus*	Scorpaenidae	Rockfish	P
Rockfish	Rockfish, black & yellow	*Sebastes chrysomelas*	Scorpaenidae	Rockfish	P
Rockfish	Rockfish, dusky	*Sebastes ciliatus*	Scorpaenidae	Rockfish	P
Rockfish	Rockfish, starry	*Sebastes constellatus*	Scorpaenidae	Rockfish	P
Rockfish	Rockfish, darkblotched	*Sebastes crameri*	Scorpaenidae	Rockfish	P
Rockfish	Rockfish, calico	*Sebastes dalli*	Scorpaenidae	Rockfish	P
Rockfish	Rockfish, splitnose	*Sebastes diploproa*	Scorpaenidae	Rockfish	P
Rockfish	Rockfish, greenstriped	*Sebastes elongatus*	Scorpaenidae	Rockfish	P
Rockfish	Rockfish, Puget Sound	*Sebastes emphaeus*	Scorpaenidae	Rockfish	P
Rockfish	Rockfish, swordspine	*Sebastes ensifer*	Scorpaenidae	Rockfish	P
Rockfish	Rockfish, widow	*Sebastes entomelas*	Scorpaenidae	Rockfish	P
Rockfish	Rockfish, pink	*Sebastes eos*	Scorpaenidae	Rockfish	P
Rockfish	Rockfish, yellowtail	*Sebastes flavidus*	Scorpaenidae	Rockfish	P
Rockfish	Rockfish, bronzespotted	*Sebastes gilli*	Scorpaenidae	Rockfish	P
Rockfish	Chilipepper	*Sebastes goodei*	Scorpaenidae	Rockfish	P
Rockfish	Rockfish, rosethorn	*Sebastes helvomaculatus*	Scorpaenidae	Rockfish	P
Rockfish	Rockfish, squarespot	*Sebastes hopkinsi*	Scorpaenidae	Rockfish	P
Rockfish	Rockfish, shortbelly	*Sebastes jordani*	Scorpaenidae	Rockfish	P
Rockfish	Rockfish, freckled	*Sebastes lentiginosus*	Scorpaenidae	Rockfish	P
Rockfish	Cowcod	*Sebastes levis*	Scorpaenidae	Rockfish	P
Rockfish	Rockfish, Mexican	*Sebastes macdonaldi*	Scorpaenidae	Rockfish	P
Rockfish	Rockfish, quillback	*Sebastes maliger*	Scorpaenidae	Rockfish	P
Rockfish	Rockfish, black	*Sebastes melanops*	Scorpaenidae	Rockfish	P
Rockfish	Rockfish, semaphore	*Sebastes melanosema*	Scorpaenidae	Rockfish	P
Rockfish	Rockfish, blackgill	*Sebastes melanostomus*	Scorpaenidae	Rockfish	P

Market Name	Common Name	Scientific Name	Family (Latin)	Family (English)	Location
Rockfish	Rockfish, vermillion	*Sebastes miniatus*	Scorpaenidae	Rockfish	P
Rockfish	Rockfish, blue	*Sebastes mystinus*	Scorpaenidae	Rockfish	P
Rockfish	Rockfish, china	*Sebastes nebulosus*	Scorpaenidae	Rockfish	P
Rockfish	Rockfish, tiger	*Sebastes nigrocinctus*	Scorpaenidae	Rockfish	P
Rockfish	Rockfish, speckled	*Sebastes ovalis*	Scorpaenidae	Rockfish	P
Rockfish	Bocaccio	*Sebastes paucispinis*	Scorpaenidae	Rockfish	P
Rockfish	Rockfish, chameleon	*Sebastes phillipsi*	Scorpaenidae	Rockfish	P
Rockfish	Rockfish, canary	*Sebastes pinniger*	Scorpaenidae	Rockfish	P
Rockfish	Rockfish, northern	*Sebastes polyspinis*	Scorpaenidae	Rockfish	P
Rockfish	Rockfish, redstripe	*Sebastes proriger*	Scorpaenidae	Rockfish	P
Rockfish	Rockfish, grass	*Sebastes rastrelliger*	Scorpaenidae	Rockfish	P
Rockfish	Rockfish, yellowmouth	*Sebastes reedi*	Scorpaenidae	Rockfish	P
Rockfish	Rockfish, rosy	*Sebastes rosaceus*	Scorpaenidae	Rockfish	P
Rockfish	Rockfish, greenblotched	*Sebastes rosenblatti*	Scorpaenidae	Rockfish	P
Rockfish	Rockfish, yelloweye	*Sebastes ruberrimus*	Scorpaenidae	Rockfish	P
Rockfish	Rockfish, flag	*Sebastes rubrivinctus*	Scorpaenidae	Rockfish	P
Rockfish	Rockfish, dwarf-red	*Sebastes rufinanus*	Scorpaenidae	Rockfish	P
Rockfish	Rockfish, bank	*Sebastes rufus*	Scorpaenidae	Rockfish	P
Rockfish	Rockfish, stripetail	*Sebastes saxicola*	Scorpaenidae	Rockfish	P
Rockfish	Rockfish, halfbanded	*Sebastes semicinctus*	Scorpaenidae	Rockfish	P
Rockfish	Rockfish, olive	*Sebastes serranoides*	Scorpaenidae	Rockfish	P
Rockfish	Treefish	*Sebastes serriceps*	Scorpaenidae	Rockfish	P
Rockfish	Rockfish, pinkrose	*Sebastes simulator*	Scorpaenidae	Rockfish	P
Rockfish	Rockfish, honeycomb	*Sebastes umbrosus*	Scorpaenidae	Rockfish	P
Rockfish	Rockfish, pygmy	*Sebastes wilsoni*	Scorpaenidae	Rockfish	P
Rockfish	Rockfish, sharpchin	*Sebastes zacentrus*	Scorpaenidae	Rockfish	P
Rockling	Rockling, fourbeard	*Enchelyopus cimbrius*	Gadidae	Codfish	A
Rosefish	Rosefish, blackbelly	*Helicolenus dactylopterus*	Scorpaenidae	Rockfish	A

A = Atlantic, P = Pacific, F = fresh water

The Fish List: Approved Names for Fish

Market Name	Common Name	Scientific Name	Family (Latin)	Family (English)	Location
Roughy	Roughy, silver	*Hoplostethus mediterraneus*	Trachichthyidae	Roughy	A-P
Roughy	Roughy, pinkfinned	*Paratrachichthys trailli*	Trachichthyidae	Roughy	P
Roughy, orange	Roughy, orange	*Hoplostethus atlanticus*	Trachichthyidae	Roughy	A-P
Rudderfish	Rudderfish	*Kyphosus cinerascens*	Kyphosidae	Chub sea	P
Ruff black	Ruff, black	*Centrolophus niger*	Stromateidae	Butterfish	P
Sablefish	Sablefish	*Anoplopoma fimbria*	Anoplopomatidae	Sablefish	P
Sailfish	Sailfish, Indo-Pacific	*Istiophorus gladius*	Istiophoridae	Billfish	A-P
Sailfish	Sailfish	*Istiophorus platypterus*	Istiophoridae	Billfish	A-P
Salmon sockeye/blueback/red	Salmon, sockeye	*Oncorhynchus nerka*	Salmonidae	Trout	P-F
Salmon, cherry	Salmon, cherry	*Oncorhynchus masou*	Salmonidae	Trout	P-F
Salmon, chinook/king/spring	Salmon, chinook	*Oncorhynchus tshawytscha*	Salmonidae	Trout	P-F
Salmon, chum/keta	Salmon, chum	*Oncorhynchus keta*	Salmonidae	Trout	P-F
Salmon, coho/silver/med.red	Salmon, coho	*Oncorhynchus kisutch*	Salmonidae	Trout	A-P-F
Salmon, pink/humpback sal.	Salmon, pink	*Oncorhynchus gorbuscha*	Salmonidae	Trout	A-F-P
Salmon/Atlantic salmon	Salmon, Atlantic	*Salmo salar*	Salmonidae	Trout	A-F
Sanddab	Sanddab, Pacific	*Citharichthys sordidus*	Bothidae	Flatfish	P
Sandlance	Sandlance, Pacific	*Ammodytes hexapterus*	Ammodytidae	Sand lance	A-P
Sandperch	Weever	*Parapercis spp.*	Mugiloididae	Sandperch	A-P
Sardine	Pilchard, false	*Harengula clupeola*	Clupeidae	Herring	A
Sardine	Sardine, redear	*Harengula humeralis*	Clupeidae	Herring	A
Sardine	Sardine, scaled	*Harengula jaguana*	Clupeidae	Herring	A-F
Sardine	Sardine, Spanish	*Sardinella anchovia*	Clupeidae	Herring	A
Sardine	Sardine, Spanish	*Sardinella aurita*	Clupeidae	Herring	A
Sardine	Sardine, orangespot	*Sardinella brasiliensis*	Clupeidae	Herring	A
Sardine	Sardine, fringescale	*Sardinella fimbriata*	Clupeidae	Herring	P
Sardine	Sardine, oil	*Sardinella longiceps*	Clupeidae	Herring	P
Sardine	Sardine, perforated-scale	*Sardinella perforata*	Clupeidae	Herring	P
Sardine/pilchard	Sardine, Pacific	*Sardinella sagax*	Clupeidae	Herring	P
Sauger	Sauger	*Stizostedion canadense*	Percidae	Perch	F

248

Market Name	Common Name	Scientific Name	Family (Latin)	Family (English)	Location
Saury	Saury, Pacific	Cololabis saira	Scomberesocidae	Saurie	P
Saury	Saury, Atlantic	Scomberesox saurus	Scomberesocidae	Saurie	A
Sawfish	Sawfish, smalltooth	Pristis pectinata	Pristidae	Sawfish	A
Sawfish	Sawfish, largetooth	Pristis perotteti	Pristidae	Sawfish	A
Scad	Scad, yellowtail	Caranx mate	Carangidae	Jack	P
Scad	Koheru	Decapterus koheru	Carangidae	Jack	P
Scad	Scad, mackerel	Decapterus maccarellus	Carangidae	Jack	A
Scad	Scad, amberstripe	Decapterus muroadsi	Carangidae	Jack	P
Scad	Scad, round	Decapterus punctatus	Carangidae	Jack	A
Scad	Scad, northern mackerel	Decapterus russelli	Carangidae	Jack	P
Scad	Scad, Mexican	Decapterus scombrinus	Carangidae	Jack	P
Scad	Scad, redtail	Decapterus tabl	Carangidae	Jack	A
Scad	Scad, bigeye	Selar crumenophthalmus	Carangidae	Jack	A
Scad	Scad, blue	Trachurus picturatus	Carangidae	Jack	A
Scad	Horse mackerel, Atlantic	Trachurus trachurus	Carangidae	Jack	A-P
Scamp	Scamp	Mycteroperca phenax	Serranidae	Seabass	A
Schoolmaster	Schoolmaster	Lutjanus apodus	Lutjanidae	Snapper	A
Scorpionfish	Scorpionfish, orange	Scorpaena scrofa	Scorpaenidae	Rockfish	A
Sculpin	Sea raven, Atlantic	Hemitripterus americanus	Cottidae	Sculpin	A
Sculpin	Sculpin, great	Myoxocephalus polyacanthocephalus	Cottidae	Sculpin	P
Sculpin/cabezon	Cabezon	Scorpaenichthys marmoratus	Cottidae	Sculpin	P
Seabream	Bream, sea	Archosargus rhomboidalis	Sparidae	Porgie	A
Seabream	Seabream	Pagellus spp.	Sparidae	Porgie	A
Searobin	Searobin, bluefin	Chelidonichthys kumu	Triglidae	Searobin	P
Searobin	Gurnard, tub	Chelidonichthys lucerna	Triglidae	Searobin	A
Searobin	Searobin, armored	Peristedion miniatum	Triglidae	Searobin	A
Searobin	Searobin, northern	Prionotus carolinus	Triglidae	Searobin	A
Searobin	Gurnard, spotted	Pterygotrigla picta	Triglidae	Searobin	P

A = Atlantic, P = Pacific, F = fresh water

The Fish List: Approved Names for Fish

Market Name	Common Name	Scientific Name	Family (Latin)	Family (English)	Location
Seatrout	Seatrout, sand	*Cynoscion arenarius*	Sciaenidae	Drum	A
Seatrout	Seatrout, spotted	*Cynoscion nebulosus*	Sciaenidae	Drum	A-F
Seatrout	Seabass, white	*Cynoscion nobilis*	Sciaenidae	Drum	P
Seatrout	Seatrout, silver	*Cynoscion nothus*	Sciaenidae	Drum	A
Shad	Shad, Alabama	*Alosa alabamae*	Clupeidae	Herring	A-F
Shad	Shad, Allis	*Alosa alosa*	Clupeidae	Herring	A-F
Shad	Shad, twaite	*Alosa fallax*	Clupeidae	Herring	A-F
Shad	Shad, hickory	*Alosa mediocris*	Clupeidae	Herring	A-F
Shad	Shad, American	*Alosa sapidissima*	Clupeidae	Herring	A-F
Shad	Shad, gizzard	*Dorosoma cepedianum*	Clupeidae	Herring	A-F
Shad	Shad, threadfin	*Dorosoma petenense*	Clupeidae	Herring	A-P-F
Shark	Shark, silver tip	*Carcharhinus albimarginatus*	Carcharhinidae	Shark	P
Shark	Shark, bignose	*Carcharhinus altimus*	Carcharhinidae	Shark	A
Shark	Shark, gray reef	*Carcharhinus amblyrhynchus*	Carcharhinidae	Shark	P
Shark	Shark, narrowtoothed	*Carcharhinus brachyurus*	Carcharhinidae	Shark	P
Shark	Shark, spinner	*Carcharhinus brevipinna*	Carcharhinidae	Shark	A
Shark	Shark, silky	*Carcharhinus falciformis*	Carcharhinidae	Shark	A
Shark	Shark, finetooth	*Carcharhinus isodon*	Carcharhinidae	Shark	A
Shark	Shark, bull	*Carcharhinus leucas*	Carcharhinidae	Shark	A-F-P
Shark	Shark, blacktip	*Carcharhinus limbatus*	Carcharhinidae	Shark	A
Shark	Shark, whitetip	*Carcharhinus longimanus*	Carcharhinidae	Shark	A-P
Shark	Shark, blacktip reef	*Carcharhinus melanopterus*	Carcharhinidae	Shark	A-P
Shark	Shark, dusky	*Carcharhinus obscurus*	Carcharhinidae	Shark	A-P
Shark	Shark, reef	*Carcharhinus perezi*	Carcharhinidae	Shark	A
Shark	Shark, sandbar	*Carcharhinus plumbeus*	Carcharhinidae	Shark	A
Shark	Shark, smalltail	*Carcharhinus porosus*	Carcharhinidae	Shark	A
Shark	Shark, night	*Carcharhinus signatus*	Carcharhinidae	Shark	A
Shark	Shark, white	*Carcharodon carcharias*	Lamnidae	Shark	A-P
Shark	Shark, basking	*Cetorhinus maximus*	Lamnidae	Shark	A-P

Market Name	Common Name	Scientific Name	Family (Latin)	Family (English)	Location
Shark	Shark, tiger	Galeocerdo cuviere	Carcharhinidae	Shark	A-P
Shark	Shark, tope	Galeorhinus galeus	Carcharhinidae	Shark	A
Shark	Shark, sixgill	Hexanchus griseus	Hexanchidae	Shark	A-P
Shark	Shark, salmon	Lamna ditropis	Lamnidae	Shark	P
Shark	Shark, lemon	Negaprion brevirostris	Carcharhinidae	Shark	A-P
Shark	Shark, sevengill	Notorynchus cepedianus	Hexanchidae	Shark	A-P
Shark	Shark, blue	Prionace glauca	Carcharhinidae	Shark	A-P
Shark	Shark, whitetip reef	Triaenodon obesus	Carcharhinidae	Shark	A-P
Shark	Shark, leopard	Triakis semifasciata	Carcharhinidae	Shark	P
Shark, mako	Shark, shortfin mako	Isurus oxyrinchus	Lamnidae	Shark	A-P
Shark, angel	Shark, Pacific angel	Squatina californica	Squatinidae	Shark	P
Shark, angel	Shark, Atlantic angel	Squatina dumerili	Squatinidae	Shark	A
Shark, mako	Shark, longfin mako	Isurus paucus	Lamnidae	Shark	A-P
Shark, thresher	Shark, pelagic thresher	Alopias pelagicus	Alopiidae	Shark	P
Shark, thresher	Shark, bigeye thresher	Alopias superciliosus	Alopiidae	Shark	A-P
Shark, thresher	Shark, common thresher	Alopias vulpinus	Alopiidae	Shark	A-P
Shark/bonnethead	Shark, bonnethead	Sphyrna tiburo	Sphyrnidae	Shark	A-P
Shark/hammerhead	Shark, scalloped hammerhead	Sphyrna lewini	Sphyrnidae	Shark	A-P
Shark/hammerhead	Shark, great hammerhead	Sphyrna mokarran	Sphyrnidae	Shark	A-P
Shark/hammerhead	Shark, smalleye hammerhead	Sphyrna tudes	Sphyrnidae	Shark	A
Shark/hammerhead	Shark, smooth hammerhead	Sphyrna zygaena	Sphyrnidae	Shark	A-P
Shark/porbeagle	Shark, porbeagle	Lamna nasus	Lamnidae	Shark	A
Shark/smoothhound	Smoothhound, grey	Mustelus californicus	Carcharhinidae	Shark	P
Shark/smoothhound	Smoothhound, brown	Mustelus henlei	Carcharhinidae	Shark	P
Sheephead	Sheephead, California	Semicossyphus pulcher	Labridae	Wrasse	P
Sheepshead	Sheepshead	Archosargus probatocephalus	Sparidae	Porgie	A-F
Silverside	Topsmelt	Atherinopsis affinis	Atherinidae	Silverside	P
Silverside	Jacksmelt	Atherinopsis californiensis	Atherinidae	Silverside	P

A = Atlantic, P = Pacific, F = fresh water

The Fish List: Approved Names for Fish

Market Name	Common Name	Scientific Name	Family (Latin)	Family (English)	Location
Silverside	Silverside, S. American	Basilichthys australis	Atherinidae	Silverside	A-F
Silverside	Silverside, Atlantic	Menidia menidia	Atherinidae	Silverside	A
Skate	Skate, Aleutian	Raja aleutica	Rajidae	Skate	P
Skate	Skate, big	Raja binoculata	Rajidae	Skate	P
Skate	Skate, clearnose	Raja eglanteria	Rajidae	Skate	A
Skate	Skate, little	Raja erinacea	Rajidae	Skate	A
Skate	Skate, rosette	Raja garmani	Rajidae	Skate	A
Skate	Skate, California	Raja inornata	Rajidae	Skate	P
Skate	Skate, Bering	Raja interrupta	Rajidae	Skate	P
Skate	Skate, sandpaper	Raja kincaidi	Rajidae	Skate	A
Skate	Skate, barndoor	Raja laevis	Rajidae	Skate	A
Skate	Skate, winter	Raja ocellata	Rajidae	Skate	A
Skate	Skate, spreadfin	Raja olseni	Rajidae	Skate	A
Skate	Skate, Alaska	Raja parmifera	Rajidae	Skate	P
Skate	Skate, thorny	Raja radiata	Rajidae	Skate	A
Skate	Skate, longnose	Raja rhina	Rajidae	Skate	P
Skate	Skate, flathead	Raja rosispinis	Rajidae	Skate	P
Skate	Skate, smooth	Raja senta	Rajidae	Skate	A
Skate	Skate, spinytail	Raja spinicauda	Rajidae	Skate	A
Skate	Skate	Raja spp.	Rajidae	Skate	A-P
Skate	Skate, starry	Raja stellulata	Rajidae	Skate	P
Skate	Skate, roundel	Raja texana	Rajidae	Skate	A
Skate	Skate, roughtail	Raja trachura	Rajidae	Skate	P
Skilfish	Skilfish	Erilepis zonifer	Anoplopomatidae	Sablefish	P
Smelt	Smelt, whitebait	Allosmerus elongatus	Osmeridae	Smelt	P
Smelt	Smelt, deep sea	Argentina semifasciata	Argentinidae	Smelt herring	P
Smelt	Smelt, great silver	Argentina silus	Argentinidae	Smelt herring	A
Smelt	Smelt, lesser silver	Argentina sphyraena	Argentinidae	Smelt herring	A
Smelt	Smelt, surf	Hypomesus pretiosus	Osmeridae	Smelt	P-F

Market Name	Common Name	Scientific Name	Family (Latin)	Family (English)	Location
Smelt	Smelt, delta	*Hypomesus transpacificus*	Osmeridae	Smelt	P-F
Smelt	Smelt, Arctic	*Osmerus dentex*	Osmeridae	Smelt	P-F
Smelt	Smelt, European	*Osmerus eperlanus*	Osmeridae	Smelt	A
Smelt	Smelt, common	*Retropinna retropinna*	Retropinnidae	Smelt southern	P
Smelt	Smelt, night	*Spirinchus starksi*	Osmeridae	Smelt	P
Smelt	Smelt, longfin	*Spirinchus thaleichthys*	Osmeridae	Smelt	P-F
Smelt	Smelt, eulachon	*Thaleichthys pacificus*	Osmeridae	Smelt	P-F
Smelt/American smelt	Smelt, rainbow	*Osmerus mordax*	Osmeridae	Smelt	A-F-P
Snake eel/keoghfish	Snake eel, giant	*Ophichthus rex*	Ophichthidae	Snake eel	A
Snapper, Caribbean red	Snapper, Caribbean red	*Lutjanus purpureus*	Lutjanidae	Snapper	A
Snapper, Pacific	Snapper, Pacific	*Lutjanus peru*	Lutjanidae	Snapper	P
Snapper, Pacific dog	Snapper, Pacific dog	*Lutjanus novemfasciatus*	Lutjanidae	Snapper	P
Snapper, black	Snapper, black	*Apsilus dentatus*	Lutjanidae	Snapper	A
Snapper, black and white	Snapper, black and white	*Macolor niger*	Lutjanidae	Snapper	P
Snapper, blackfin	Snapper, blackfin	*Lutjanus buccanella*	Lutjanidae	Snapper	A
Snapper, blacktail	Snapper, blacktail	*Lutjanus fulvus*	Lutjanidae	Snapper	P
Snapper, blubberlip	Snapper, blubberlip	*Lutjanus rivulatus*	Lutjanidae	Snapper	P
Snapper, bluestriped	Snapper, bluestriped	*Lutjanus kasmira*	Lutjanidae	Snapper	P
Snapper, cardinal	Snapper, cardinal	*Pristipomoides macrophthalmus*	Lutjanidae	Snapper	A
Snapper, colorado	Snapper, colorado	*Lutjanus colorado*	Lutjanidae	Snapper	P
Snapper, crimson	Snapper, crimson	*Pristipomoides filamentosus*	Lutjanidae	Snapper	P
Snapper, cubera	Snapper, cubera	*Lutjanus cyanopterus*	Lutjanidae	Snapper	A
Snapper, dog	Snapper, dog	*Lutjanus jocu*	Lutjanidae	Snapper	A
Snapper, emperor	Snapper, emperor	*Lutjanus sebae*	Lutjanidae	Snapper	P
Snapper, golden	Snapper, golden	*Lutjanus inermis*	Lutjanidae	Snapper	P
Snapper, gray	Snapper, gray	*Lutjanus griseus*	Lutjanidae	Snapper	A-F
Snapper, humpback	Snapper, humpback	*Lutjanus gibbus*	Lutjanidae	Snapper	P
Snapper, lane	Snapper, lane	*Lutjanus synagris*	Lutjanidae	Snapper	A

A = Atlantic, P = Pacific, F = fresh water

The Fish List: Approved Names for Fish

Market Name	Common Name	Scientific Name	Family (Latin)	Family (English)	Location
Snapper, mahogony	Snapper, mahogany	*Lutjanus mahogoni*	Lutjanidae	Snapper	A
Snapper, malabar	Malabar	*Lutjanus malabaricus*	Lutjanidae	Snapper	P
Snapper, midnight	Snapper, midnight	*Macolor macularius*	Lutjanidae	Snapper	P
Snapper, mullet	Snapper, mullet	*Lutjanus aratus*	Lutjanidae	Snapper	P
Snapper, mutton	Snapper, mutton	*Lutjanus analis*	Lutjanidae	Snapper	A
Snapper, onespot	Snapper, onespot	*Lutjanus monostigma*	Lutjanidae	Snapper	P
Snapper, queen	Snapper, queen	*Etelis oculatus*	Lutjanidae	Snapper	A
Snapper, red	Snapper, red	*Lutjanus campechanus*	Lutjanidae	Snapper	A
Snapper, ruby	Snapper, ruby	*Etelis carbunculus*	Lutjanidae	Snapper	P
Snapper, rufous	Snapper, rufous	*Lutjanus jordani*	Lutjanidae	Snapper	P
Snapper, sailfin	Snapper, sailfin	*Symphorichthys spilurus*	Lutjanidae	Snapper	P
Snapper, scarlet	Snapper, scarlet	*Lutjanus sanguineus*	Lutjanidae	Snapper	P
Snapper, silk	Snapper, silk	*Lutjanus vivanus*	Lutjanidae	Snapper	A
Snapper, spotted rose	Snapper, spotted rose	*Lutjanus guttatus*	Lutjanidae	Snapper	P
Snapper, twinspot	Snapper, twinspot	*Lutjanus bohar*	Lutjanidae	Snapper	P
Snapper, vermilion	Snapper, vermilion	*Rhomboplites aurorubens*	Lutjanidae	Snapper	A
Snapper, yellowstriped	Snapper, yellowstriped	*Etelis coruscans*	Lutjanidae	Snapper	P
Snapper, yellowtail	Snapper, yellowtail	*Ocyurus chrysurus*	Lutjanidae	Snapper	A
Snook	Snook, swordspine	*Centropomus ensiferus*	Centropomidae	Snook	A-F
Snook	Snook	*Centropomus nigrescens*	Centropomidae	Snook	P-F
Snook	Snook	*Centropomus ornatus*	Centropomidae	Snook	P-F
Snook	Snook, fat	*Centropomus parallelus*	Centropomidae	Snook	A-F
Snook	Snook, tarpon	*Centropomus pectinatus*	Centropomidae	Snook	A-F
Snook	Snook	*Centropomus pedemacula*	Centropomidae	Snook	P-F
Snook	Constintino	*Centropomus robalito*	Centropomidae	Snook	P-F
Snook	Snook	*Centropomus undecimalis*	Centropomidae	Snook	A-F
Snook	Snook	*Centropomus vindis*	Centropomidae	Snook	P-F
Sole	Sole, kobe	*Aseraggodes kobensis*	Soleidae	Flatfish	P
Sole	Sole, narrowbanded	*Aseraggodes macleayanus*	Soleidae	Flatfish	P

254

Market Name	Common Name	Scientific Name	Family (Latin)	Family (English)	Location
Sole	Sole	*Austroglossus microlepis*	Soleidae	Flatfish	A
Sole	Sole	*Austroglossus pectoralis*	Soleidae	Flatfish	A
Sole	Sole, slender	*Lyopsetta exilis*	Pleuronectidae	Flatfish	P
Sole	Sole, thickback	*Microchirus variegatus*	Soleidae	Flatfish	A
Sole	Sole, lemon	*Microstomus kitt*	Pleuronectidae	Flatfish	A
Sole	Sole, English	*Parophrys vetulus*	Pleuronectidae	Flatfish	P
Sole	Sole, European	*Solea vulgaris*	Soleidae	Flatfish	A
Sole	Sole, oriental black	*Synaptura orientalis*	Soleidae	Flatfish	A
Sole dover	Sole, dover	*Microstomus pacificus*	Pleuronectidae	Flatfish	P
Sole/flounder	Sole, roughscale	*Clidoderma asperrimum*	Pleuronectidae	Flatfish	P
Sole/flounder	Sole, deepsea	*Embassichthys bathybius*	Pleuronectidae	Flatfish	P
Sole/flounder	Sole, petrale	*Eopsetta jordani*	Pleuronectidae	Flatfish	P
Sole/flounder	Flounder, witch/gray sole	*Glyptocephalus cynoglossus*	Pleuronectidae	Flatfish	A
Sole/flounder	Sole, rex	*Glyptocephalus zachirus*	Pleuronectidae	Flatfish	P
Sole/flounder	Sole, bigmouth	*Hippoglossina stomata*	Pleuronectidae	Flatfish	P
Sole/flounder	Sole, flathead	*Hippoglossoides elassodon*	Bothidae	Flatfish	P
Sole/flounder	Sole, butter	*Isopsetta isolepis*	Pleuronectidae	Flatfish	P
Sole/flounder	Sole, rock	*Lepidopsetta bilineata*	Pleuronectidae	Flatfish	P
Sole/flounder	Sole, yellowfin	*Limanda aspera*	Pleuronectidae	Flatfish	P
Sole/flounder	Sole, C-O	*Pleuronichthys coenosus*	Pleuronectidae	Flatfish	P
Sole/flounder	Sole, curlfin	*Pleuronichthys decurrens*	Pleuronectidae	Flatfish	P
Sole/flounder	Sole, sand	*Psettichthys melanosticus*	Pleuronectidae	Flatfish	P
Sole/flounder	Sole, fantail	*Xystreurys liolepis*	Bothidae	Flatfish	P
Spadefish	Spadefish, Atlantic	*Chaetodipterus faber*	Ephippidae	Spadefish	A
Spadefish	Spadefish, Pacific	*Chaetodipterus zonatus*	Ephippidae	Spadefish	P
Spearfish	Spearfish, shortbill	*Tetrapturus angustirostris*	Istiophoridae	Billfish	P
Spearfish	Spearfish, longbill	*Tetrapturus pfluegeri*	Istiophoridae	Billfish	A
Spinefoot	Spinefoot	*Siganus spp.*	Siganidae	Rabbitfish	A-P

A = Atlantic, P = Pacific, F = fresh water

The Fish List: Approved Names for Fish

Market Name	Common Name	Scientific Name	Family (Latin)	Family (English)	Location
Spot	Spot	*Leiostomus xanthurus*	Sciaenidae	Drum	A-F
Sprat/brisling	Sprat	*Sprattus spp.*	Clupeidae	Herring	A-P
Squirrelfish	Squirrelfish	*Holocentrus ascensionis*	Holocentridae	Squirrelfish	A
Squirrelfish	Squirrelfish, longspine	*Holocentrus rufus*	Holocentridae	Squirrelfish	A
Squirrelfish	Squirrelfish	*Myripristis argyromus*	Holocentridae	Squirrelfish	P
Squirrelfish	Squirrelfish	*Myripristis berndti*	Holocentridae	Squirrelfish	P
Squirrelfish	Squirrelfish	*Sargocentron lacteoguttatum*	Holocentridae	Squirrelfish	P
Squirrelfish	Squirrelfish, scarlet	*Sargocentron spiniferum*	Holocentridae	Squirrelfish	P
Stargazer	Stargazer, spotted	*Genyagnus monopterygius*	Uranoscopidae	Stargazer	P
Stargazer	Stargazer, smooth	*Kathetostoma averruncus*	Uranoscopidae	Stargazer	P
Stargazer	Stargazer, giant	*Kathetostoma giganteum*	Uranoscopidae	Stargazer	P
Sturgeon	Sturgeon, Russian	*Acipenser gueldenstaedti*	Acipenseridae	Sturgeon	F
Sturgeon	Sturgeon, green	*Acipenser medirostris*	Acipenseridae	Sturgeon	P-F
Sturgeon	Sturgeon, Atlantic	*Acipenser oxyrhynchus*	Acipenseridae	Sturgeon	A-F
Sturgeon	Sturgeon, star	*Acipenser stellatus*	Acipenseridae	Sturgeon	F
Sturgeon	Sturgeon, white	*Acipenser transmontanus*	Acipenseridae	Sturgeon	P-F
Sturgeon	Sturgeon, European	*Huso huso*	Acipenseridae	Sturgeon	F
Sucker	Carpsucker, river	*Carpiodes carpio*	Catostomidae	Sucker	F
Sucker	Quillback	*Carpiodes cyprinus*	Catostomidae	Sucker	F
Sucker	Sucker, white	*Catostomus commersoni*	Catostomidae	Sucker	F
Sucker	Sucker, blue	*Cycleptus elongatus*	Catostomidae	Sucker	F
Sucker/redhorse	Redhorse, shorthead	*Moxostoma macrolepidotum*	Catostomidae	Sucker	F
Sunfish	Perch, Sacramento	*Archoplites interruptus*	Centrarchidae	Sunfish	F
Sunfish	Sunfish, redbreast	*Lepomis auritus*	Centrarchidae	Sunfish	F
Sunfish	Sunfish, green	*Lepomis cyanellus*	Centrarchidae	Sunfish	F
Sunfish	Pumpkinseed	*Lepomis gibbosus*	Centrarchidae	Sunfish	F
Sunfish	Sunfish, redear	*Lepomis microlophus*	Centrarchidae	Sunfish	F
Sunfish	Sunfish, spotted	*Lepomis punctatus*	Centrarchidae	Sunfish	F
Surfperch	Surfperch, barred	*Amphistichus argenteus*	Embiotocidae	Surfperch	P

Market Name	Common Name	Scientific Name	Family (Latin)	Family (English)	Location
Surfperch	Surfperch, calico	Amphistichus koelzi	Embiotocidae	Surfperch	P
Surfperch	Surfperch, redtail	Amphistichus rhodoterus	Embiotocidae	Surfperch	P
Surfperch	Surfperch, shiner	Cymatogaster aggregata	Embiotocidae	Surfperch	P-F
Surfperch	Surfperch, black	Embiotoca jacksoni	Embiotocidae	Surfperch	P
Surfperch	Surfperch, striped	Embiotoca lateralis	Embiotocidae	Surfperch	P
Surfperch	Surfperch, walleye	Hypersopon argenteum	Embiotocidae	Surfperch	P
Surfperch	Seaperch, rubberlip	Rhacochilus toxotes	Embiotocidae	Surfperch	P
Swordfish	Swordfish	Xiphias gladius	Xiphiidae	Swordfish	A-P
Tang	Doctorfish	Acanthurus chirurgus	Acanthuridae	Surgeonfish	A
Tang	Tang, blue	Acanthurus coeruleus	Acanthuridae	Surgeonfish	A
Tarpon	Tarpon, Atlantic	Megalops atlanticus	Elopidae	Tarpon	A-F
Tautog	Tautog	Tautoga onitis	Labridae	Wrasse	A
Thornyhead	Thornyhead, shortspine	Sebastolobus alascanus	Scorpaenidae	Rockfish	P
Thornyhead	Thornyhead, longspine	Sebastolobus altivelis	Scorpaenidae	Rockfish	P
Tilapia	Tilapia, blue	Tilapia aurea	Cichlidae	Cichlid	F
Tilapai	Tilapia, mango	Tilapia galilaea	Cichlidae	Cichlid	F
Tilapia	Tilapia, longfin	Tilapia macrochir	Cichlidae	Cichlid	F
Tilapia	Tilapia, blackchin	Tilapia melanotheron	Cichlidae	Cichlid	F
Tilapia	Tilapia, Mozambique	Tilapia mossambica	Cichlidae	Cichlid	F
Tilapia	Tilapia, Nile	Tilapia nilotica	Cichlidae	Cichlid	F
Tilapia	Tilapia, redbreast	Tilapia rendalli	Cichlidae	Cichlid	F
Tilefish	Tilefish, goldface	Caulolatilus chrysops	Malacanthidae	Tilefish	A
Tilefish	Tilefish, blueline	Caulolatilus microps	Malacanthidae	Tilefish	A
Tilefish	Whitefish, ocean	Caulolatilus princeps	Malacanthidae	Tilefish	P
Tilefish	Tilefish, golden	Lopholatilus chamaeleonticeps	Malacanthidae	Tilefish	A
Tilefish	Tilefish, sand	Malacanthus plumieri	Malacanthidae	Tilefish	A
Toadfish	Toadfish, Gulf	Opsanus beta	Batrachoididae	Toadfish	A
Toadfish	Toadfish, oyster	Opsanus tau	Batrachoididae	Toadfish	A

A = Atlantic, P = Pacific, F = fresh water

The Fish List: Approved Names for Fish

Market Name	Common Name	Scientific Name	Family (Latin)	Family (English)	Location
Tomcod	Tomcod, Pacific	Microgadus proximus	Gadidae	Codfish	P
Tomcod	Tomcod, Atlantic	Microgadus tomcod	Gadidae	Codfish	A-F
Tonguesole	Tonguesole	Cynoglossus spp.	Cynoglossidae	Flatfish	P
Torpedo	Torpedo, Atlantic	Torpedo nobiliana	Torpedinidae	Ray electric	A
Trevally	Trevally, white	Caranx dentex	Carangidae	Jack	P
Trevally	Trevally, bigeye	Caranx sexfasciatus	Carangidae	Jack	P
Triggerfish	Triggerfish, gray	Balistes capriscus	Balistidae	Leatherjacket	A
Triggerfish	Triggerfish, queen	Balistes vetula	Balistidae	Leatherjacket	A
Triggerfish	Triggerfish, ocean	Cantherdermis sufflamen	Balistidae	Leatherjacket	A
Triggerfish	Filefish, fringed	Melichthys niger	Balistidae	Leatherjacket	A-P
Triggerfish	Triggerfish	Navodon convexirostris	Balistidae	Leatherjacket	P
Triggerfish	Triggerfish	Navodon scabra	Balistidae	Leatherjacket	P
Tripletail	Tripletail, fourband	Datinoides quadrifasciatus	Lobotidae	Tripletail	P
Tripletail	Tripletail, west coast	Lobotes pacificus	Lobotidae	Tripletail	P
Tripletail	Tripletail	Lobotes surinamensis	Lobotidae	Tripletail	A
Trout	Trout, golden	Salmo aguabonita	Salmonidae	Trout	F
Trout	Trout, gila	Salmo gilae	Salmonidae	Trout	F
Trout	Trout, brown	Salmo trutta	Salmonidae	Trout	A-F
Trout, brook	Trout, brook	Salvelinus fontinalis	Salmonidae	Trout	A-F
Trout, cutthroat	Trout, cutthroat	Salmo clarki	Salmonidae	Trout	P-F
Trout, lake	Trout,lake	Salvelinus namaycush	Salmonidae	Trout	F
Trout, rainbow/steelhead	Trout, rainbow	Salmo gairdneri*	Salmonidae	Trout	A-F-P
Trout/Dolly Varden	Trout, Dolly Varden	Salvelinus malma	Salmonidae	Trout	P-F
Trout/inconnu	Inconnu	Stenodus leucichthys	Salmonidae	Trout	F
Trumpeter	Trumpeter, bastard	Latridopsis ciliaris	Latrididae	Trumpeter	P
Trumpeter	Trumpeter, striped	Latris lineata	Latrididae	Trumpeter	P
Tuna	Tuna, bullet	Auxis rochei	Scombridae	Mackerel	A-P
Tuna	Kawakawa	Euthynnus affinis	Scombridae	Mackerel	P
Tuna	Tunny, little	Euthynnus alletteratus	Scombridae	Mackerel	A

Market Name	Common Name	Scientific Name	Family (Latin)	Family (English)	Location
Tuna	Skipjack, black	Euthynnus lineatus	Scombridae	Mackerel	P
Tuna	Tuna, skipjack	Euthynnus pelamis	Scombridae	Mackerel	A-P
Tuna	Albacore	Thunnus alalunga	Scombridae	Mackerel	A-P
Tuna	Tuna, yellowfin	Thunnus albacares	Scombridae	Mackerel	A-P
Tuna	Tuna, blackfin	Thunnus atlanticus	Scombridae	Mackerel	A
Tuna	Tuna, southern bluefin	Thunnus maccoyii	Scombridae	Mackerel	A-P
Tuna	Tuna, bigeye	Thunnus obesus	Scombridae	Mackerel	A-P
Tuna	Tuna, bluefin	Thunnus thynnus	Scombridae	Mackerel	A-P
Tuna	Tuna, longtail	Thunnus tonggol	Scombridae	Mackerel	P
Turbot	Turbot, diamond	Hypsopsetta guttulata	Pleuronectidae	Flatfish	P
Turbot	Turbot, spotted	Pleuronichthys ritteri	Pleuronectidae	Flatfish	P
Turbot	Turbot, hornyhead	Pleuronichthys verticalis	Pleuronectidae	Flatfish	P
Turbot	Turbot, spottedtail	Psettodes belcheni	Psettodidae	Flatfish	A
Turbot	Turbot, spring	Psettodes bennetti	Psettodidae	Flatfish	P
Turbot	Turbot	Scophthalmus maximus	Bothidae	Flatfish	A
Turbot, Greenland	Halibut, Greenland	Reinhardtius hippoglossoides	Pleuronectidae	Flatfish	A-P
Wahoo	Wahoo	Acanthocybium solanderi	Scombridae	Mackerel	A-P
Walleye	Pike, walleye	Stizostedion vitreum	Percidae	Perch	F
Warehou	Warehou, blue	Seriolella brama	Centrolophidae	Warehou	P
Warehou	Warehou, white	Seriolella caerulee	Centrolophidae	Warehou	P
Warehou	Warehou, silver	Seriolella punctata	Centrolophidae	Warehou	P
Weakfish	Weakfish	Cynoscion spp.	Sciaenidae	Drum	A-P
Weakfish	Weakfish, king	Macrodon ancylodon	Sciaenidae	Drum	A
Wenchman	Wenchman	Pristipomoides aquilonaris	Lutjanidae	Snapper	A
Whitefish	Whitefish, lake	Coregonus clupeaformis	Salmonidae	Trout	A-F
Whitefish	Whitefish, humpback	Coregonus pidschian	Salmonidae	Trout	F
Whitefish	Whitefish, round	Prosopium cylindraceum	Salmonidae	Trout	F
Whiting	Whiting, European	Merlangus merlangus	Gadidae	Codfish	A

A = Atlantic, P = Pacific, F = fresh water
* The scientific name changed 1/1/90.

The Fish List: Approved Names for Fish

Market Name	Common Name	Scientific Name	Family (Latin)	Family (English)	Location
Whiting	Hake, N.Z./Antarctic queen*	*Merluccius australis*	Gadidae	Codfish	A
Whiting	Hake, silver	*Merluccius bilinearis*	Gadidae	Codfish	A
Whiting	Hake, Cape	*Merluccius capensis*	Gadidae	Codfish	A-P
Whiting	Hake, Chilean	*Merluccius gayi*	Gadidae	Codfish	P
Whiting	Hake, Argentine	*Merluccius hubbsi*	Gadidae	Codfish	A
Whiting	Hake, European	*Merluccius merluccius*	Gadidae	Codfish	A
Whiting	Hake, Patagonian	*Merluccius polylepsis*	Gadidae	Codfish	A
Whiting New Zealand	Hoki	*Macruronus novaezealandiae*	Gadidae	Codfish	P
Whiting blue	Whiting, southern blue	*Micromesistius australis*	Gadidae	Codfish	A-P
Whiting blue	Poutassou	*Micromesistius poutassou*	Gadidae	Codfish	A
Whiting/Pacific whiting	Hake, Pacific	*Merluccius productus*	Gadidae	Codfish	P

A = Atlantic, P = Pacific, F = fresh water
* The name Antarctic queen has been approved for *Merluccius australis* since the *List* was originally published.

Index

Note: Index entries in UPPER CASE indicate headings or subordinate headings in the text of the book.

A.O.A.C., 7
ABALONE, 2
Abrupta, 33
Acanthias, 161
Acanthocybium solanderi, 202
Acipenser medirostris, 183
Acipenser spp., 182
Acipenser transmontanus, 183
Acutus, 43
ADDITIVES, 3
Adductor muscle, 149
Adscensionis, 72
Aeglefinus, 74
Aguabonita, 192
Ahi, 196
AIR FREIGHT, 4
Air Transport Association of America, 4
Aku, 196
Alalunga, 194
Alaska cod, 34, 126
Albacares, 196
ALBACORE, 194
Albula, 201
ALEWIFE, 4, 80
Alletteratus, 195
ALLIGATOR, 5
Allmouth, 106

Allocyttus spp., 50
Alosa pseudoharengus, 4
Alosa sapidissima, 158
Alpinus, 23
Alutus, 132
AMARELO, 146
AMBERJACK, 5, 24, 209
American eel, 52
American John Dory, 50
AMERICAN OYSTER, 119
American plaice, 59
Americanus, 58, 61, 88, 91, 106, 114, 125
Ammodytes spp., 147
Ammonia, 13
Anadromous, 4, 19
ANADROMOUS FISH, 6
Anaerobic, 121
ANAEROBIC BACTERIA, 6
Analis, 178
Anarhichas spp., 21
Anchovia, 148
ANCHOVY, 6, 80, 148
Andamanicus, 100
Anglerfish, 106
Anguilla, 52
Anguilla anguilla, 52
Anguilla japonicus, 52

INDEX

Anguilla rostrata, 16, 52, 136
ANTARCTIC QUEEN, 206
Anti-oxidants, 3
ANTIBIOTICS, 7, 77
Aplodinotus grunniens, 161
Apsilus dentatus, 177
AQUACULTURE, 7, 95
Archosargus probatocephalus, 161
Arctic char, 23
Arctica islandica, 31
Arenaria, 27
Arenarius, 157
Argentea, 11
Argentina, 168
Argentine, 168
Argopecten circularis, 152
Argopecten gibbus, 152
Argopecten irradians, 150
Arius felis, 21
ARK, 26, 33
ARROWTOOTH FLOUNDER, 58, 197
Aspera, 62
Assimilis, 2
Association of Official Analytical Chemists, 7
Astacus, 43
Astacus astacus, 43
Atheresthes stomias, 58
ATKA MACKEREL, 9
Atlantic barracuda, 24
Atlantic cod, 34
Atlantic halibut, 75
Atlantic jackknife clam, 32
ATLANTIC MACKEREL, 101
Atlantic oyster, 119
Atlantic pollock, 15
Atlantic razor clam, 32
ATLANTIC SALMON, 139
Atlantic saury, 111
Atlantic surfclam, 31
ATLANTIC WHITING, 206
Atlanticus, 115, 194
Aureomycin, 7
Aurita, 148

Aurorubens, 178
Australis, 206, 207
Auxis thazard, 68, 102
Azonus, 9
Aztecus, 166

Bacalao, 145, 146
Backfin, 40
Bacteria, 46, 185
Balistes spp., 190
Bar clam, 31
BARNACLES, 10
BARRACUDA, 10, 11
Barramundi, 19
Basket cockle, 33
BASS, 11
Bay crabs, 41, 150
Beards, 109
Belly, 27
BELLY BURN, 12
BELLY CLAM, 26, 27
Bellyfish, 106
Bellyflaps, 161
Beluga, 22
Bering Sea flounder, 59
BIGEYE, 194
Bigmouth buffalo, 15
Bilinearis, 206
Bilineata, 58, 61
BILLFISH, 12
Bismark herring, 81
BISQUE, 12
Bivalves, 106
Black abalone, 2
Black bass, 15
Black buffalo, 16
Black cod, 34, 72, 136
Black halibut, 75, 196
Black jewfish, 73
Black mullet, 107
Black oreo dory, 50
Black quahog, 31
BLACK SEA BASS, 11
BLACK SNAPPER, 177

INDEX

Black spot,*105, 167
Blackback, 58
Blackback flounder, 61
BLACKFIN, 194
BLACKFIN SNAPPER, 177
Blackfish, 11, 188
Blackfordii, 176
Blackmouth, 141
BLEEDING, 12
Blind robins, 173
BLOATERS, 13, 25, 172, 173, 175
Blood, 12
Blood cockles, 33
Blowfish, 128
Blowing, 118
Blue cod, 125
BLUE CRAB, 38
BLUE MUSSELS, 108
Blue pike, 124
BLUE RUNNER, 86, 122
BLUE TROUT, 192
Blueback, 142, 144
BLUEFIN, 194
Bluefin tuna, 102
BLUEFISH, 13
Bluegills, 183
Bocaccio, 132
Body meat, 40
Bonaci, 72
BONELESS COD, 15, 125
BONELESS FILLETS, 14
BONITO, 15, 102, 193
Borealis, 165
BOSTON BLUEFISH, 13, 15, 125
Bottomfish, 46
Botulinus, 4, 6, 121
Botulism, 6, 172
Brabatus, 107
Brachyplatysoma vaillanti, 19
Brama, 126
Brama brama, 126
Brama japonica, 126
BRANCO, 146
BREAM, 15
Brevirostris, 134

Brevoortia tyrannus, 105
Bright chums, 15, 143
Brook trout, 190, 192
Brosme brosme, 45
Brown eels, 52
Brown shrimp, 166
Brown tide, 151
Brown trout, 190, 192
Bruises, 12
BTS, 193
Bubalus, 15
Buccanella, 177
BUCKLING, 15, 172, 173
Buckrams, 38
BUFFALO, 15, 107
Bullheads, 19
BURBOT, 16
Burnt tuna syndrome, 193
Burti, 16
BUSHEL, 16, 26
Busycon carica, 203
Busycotypus canaliculatus, 203
BUTTER CLAM, 32
BUTTERFISH, 16, 122, 136
Byssal threads, 109

Caballus, 86
CALAMARE, 18
Calamari, 179
Calico, 143
CALICO SCALLOP, 150, 152
California barracuda, 11
California halibut, 75
California sardine, 148
CALIFORNIA SQUID, 181
Californicus, 75
Callinectes sapidus, 38
Campechanus, 176
Campechiensis, 29
Canadense, 124
Canadian Cooking Theory, 37
Canadum, 33
Canaliculatus, 203
Canaliculus, 110

INDEX

Cancer borealis, 42
Cancer edwardsii, 42
Cancer irroratus, 42
Cancer magister, 41
Candlefish, 168
Candling, 123
Canis, 161
Canned foods, 6
CAP, 122
Cape shark, 161
CAPELIN, 18, 135, 168, 169
Capelin roe, 18
Carangidae, 86
Caranx caballus, 86
Caranx crysos, 44, 86
Caranx hippos, 86
Carica, 203
Carolinus, 127, 156
Carp, 8, 18
CARPET SHELL, 19
Carpio, 18
CATADROMOUS, 19, 52
Catastomus spp., 16
Catfish, 8
CATFISH, FRESHWATER, 19
CATFISH, OCEAN, 21
CATFISH, SEA, 21
Caurinus, 153
Cavalla, 44, 103
CAVIAR, 21, 100, 135, 182
CAVITATION, 11, 23
CEPHALOPOD, 23, 106, 179
Cephalus, 107
CERO, 102
Cervimunida johni, 100
Ceviche, 158
Chain pickerel, 125
Chalcogrammus, 126
Chamaelonticeps, 189
Channel bass, 11
Channel catfish, 19
Channeled whelk, 203
Chanos, 106
Chanos chanos, 106
CHAR, 23

Charr, 23
CHERRY, 144
Cherry salmon, 137
CHERRYSTONE, 26, 27, 29, 120
Chiliensis, 15
Chilipepper, 132
Chiloe oyster, 120
CHINOOK, 141
Chinook salmon, 137
Chlamys farreri, 151
Chlamys hastata, 152
Chlamys opercularis, 151
Chlamys rubida, 152
Chlorine, 46
Cholera, 162
CHOPPED CLAMS, 26
Chopper, 13
CHOWDER, 24, 29
CHOWDER CLAM, 26, 27
Chrysops, 128, 182, 204
Chrysurus, 178
CHUB, 24, 143, 175
Chum salmon, 137
Chuss, 207
CIGUATERA, 24
Ciguatoxins, 24
Circularis, 152
CISCO, 13, 24, 175, 205
Citric acid, 49
CLAM JUICE, 26
Clam strips, 28, 31
Clambake, 28
CLAMS, 25, 46, 78
Clarki, 192
Clarkii, 43
Claw meat, 40
Clinocardium nuttallii, 33
Cluckers, 80, 86
Clupeaformis, 205
Coalfish, 125
COBIA, 33
COCKLE, 26, 33
Cocktail claws, 40
Cod, 8, 16, 34, 55, 172
Cod cheeks, 36

INDEX

Cod roes, 36
Cod tongues, 36
CODEX ALIMENTARIUS, 36
Codfish, 34
COHO, 142
Coho salmon, 137
Coley, 125
Cololabis saira, 111
CONCH, 37, 203
Concholepas spp., 3
CONGER EEL, 37
Conger oceanicus, 37, 122
CONVICT FISH, 161
Cooking, 96
COOKING FISH, 37
CORAL, 38
Coregonus albula, 201
Coregonus clupeaformis, 205
Coregonus spp., 24
Corrugata, 2
CORVINA, 38
Coryphaena hippurus, 104
Count necks, 29
Counts, 117
Cove oyster, 119
Cowcod, 132
CRABS, 12, 38, 79
Crackerodii, 2
Crappie, 15, 183
Crassostrea gigas, 119
Crassostrea rivularis, 120
Crassostrea virginica, 119
CRAWFISH, 43, 99
Crepidula fornicata, 168
Crevalle, 44
CREVALLE JACK, 44
CROAKER, 45, 122, 179
Cromis, 50
CRUSTACEANS, 45
Crysos, 86
Ctenopharyngodon idella, 18
Cuban cod, 161
CUBERA SNAPPER, 177
Cupped oyster, 119
CURED FISH, 145

Curema, 107
Cusk, 21, 45
CUTTHROAT TROUT, 192
Cuttlefish, 3, 23, 45, 83, 177
Cyclopterus lumpus, 100
Cynoglossus, 58, 60
Cynoscion arenarius, 157
Cynoscion nebulosus, 157
Cynoscion nobilis, 11
Cynoscion nothus, 157
Cynoscion parvipinnis, 38
Cynoscion regalis, 157
Cynoscion spp., 157
Cyprinellus, 15
Cyprinus carpio, 18

Dab, 58, 59
Dainty tails, 99
Dark chum, 143
Deep sea dainties, 99
Deep sea lobster, 99
Dehydration, 145
DEMERSAL, 46
Dentatus, 58, 60, 177
DEPURATION, 46
Dill herring, 81
Dinoflagellate algae, 24
DIPS, 48
Directus, 32
Discus, 2
Dog shark, 161
DOG SNAPPER, 177
DOGFISH, 161
Dollar fish, 16
Dolly Varden, 23
DOLPHIN, 49, 104
Domoic acid poisoning, 130
Dorado, 104
Doré, 124
DORIES, 50
Dover sole, 58, 59
DOVER SOLE, GENUINE, 59
DRAWN, 50
DRESSED, 50

INDEX

DRIED FISH, 145
Drum, 11, 157
DRUM, BLACK, 50
Drummondhayi, 72
DRY SALTING, 146
Dublin Bay prawn, 99
Dumerili, 5
DUNGENESS CRAB, 41
Duorarum, 166

E. coli, 162
Eastern oyster, 119
Echinoderms, 158
Edulis, 108, 120
EEL, 19, 52, 173
Eelpout, 114
Eisenbecki, 43
Electrophoresis, 85
Elongatus, 90
Elops saurus, 90
Elvers, 52
English sole, 58, 59
Engraulis mordax, 6
Ensis directus, 32
Eopsetta jordani, 58, 60
Epinephelus adscensionis, 72
Epinephelus drummondhayi, 72
Epinephelus flavolimbatus, 73
Epinephelus guttatus, 72
Epinephelus itajara, 72
Epinephelus morio, 72
Epinephelus mystacinus, 72
Epinephelus nigritus, 73
Epinephelus niveatus, 72
Epinephelus striatus, 72
Erimacrus eisenbecki, 43
Esox americanus, 125
Esox lucius, 124
Esox masquinongy, 125
Esox niger, 125
Eulachon, 168
Euphausia superba, 89
Euphausia superba, 11
European eel, 52

European John Dory, 50
Euthynnus alletteratus, 195
Euthynnus pelamis, 195
Eviscerated, 50
Extra selects, 117
Eyes, 68

Faber, 50
Fair Packaging and Labeling Act, 203
Falcatus, 127
FAO, 7, 36
Farm-raised catfish, 19
Farreri, 151
Fasciatus, 131
FAT FISH, 54
Fathead, 161
Felis, 21
Ferruginea, 58, 62
Filefish, 190
FILLET, 54
FILTER FEEDERS, 46, 54
Fimbria, 16, 136
FINNAN, 54
FINNAN HADDOCK, 173
Fish balls, 70
Fish eggs, 135
Fish List, 58, 112, 176
FISHERIES MANAGEMENT, 55
FISHING METHODS, 55
FLAKE, 56, 69, 161
Flake meat, 40, 56
Flat abalone, 2
FLAT OYSTER, 120
Flatfish, 57
Flavescens, 90
Flavolimbatus, 73
FLETCH, 56, 76
Flounder, 57, 60, 122
Fluke, 58, 60, 61, 122
Fontinelis, 192
Food and Agriculture Organization, 36
Foot, 3

INDEX

Fornicata, 168
FRAUD, 63
FRESHNESS, 65
FRESHWATER DRUM, 161
Freshwater eel, 52
FRIGATE MACKEREL, 15, 68, 102
Frill, 54
Frogfish, 106
FROGS, 68
FRYERS, 26, 27
Fugu, 128

Gadus macrocephalus, 34
Gadus macrocephalus, 34
Gag, 71, 72
Gairdneri, 141, 191
Galloprovincialis, 108
Gamma radiation, 84
Gammarus, 91
Gaper clam, 33
GAPING, 69
Gas-flushed packaging, 4
Gaspareau, 4
GASPE, 147
Gaspergoo, 4, 161
Gastropod, 37
GEFILTE FISH, 70
Generally Recognized as Safe, 3
Genypterus spp., 88
GEODUCK, 27, 33
Giant bluefin, 56, 195
Giant freshwater shrimp, 167
GIBBED, 70, 152
Giganteus, 32
Gigas, 37, 119
Gila trout, 190
Gills, 67
Gladius, 186
Glass eels, 52
Globefish, 128
Glyptocephalus cynoglossus, 58, 60
Glyptocephalus zachirus, 60

GMP, 70
GOATFISH, 70
Golden caviar, 22
Golden tilefish, 189
Golden trout, 190, 192
GONADS, 70, 135
GOOD MANUFACTURING PRACTICE, 70
Goodei, 10, 127, 132
Goose-necked barnacles, 10
Goosefish, 106
Gorbuscha, 137, 143
GRAS, 3, 71, 183
Grass carp, 18
Grass pickerel, 125
GRAVLAX, 71
Gray herring, 4
Gray mullet, 107
Gray shark, 161
GRAY SNAPPER, 178
Gray sole, 58, 60
Great barracuda, 11
Green cod, 125
Green jack, 86
Green lipped mussels, 110
GREEN MUSSELS, 110
GREEN STURGEON, 183
Greenland halibut, 75, 196
Greenling, 9, 90
GRENADIER, 71
GREY SEATROUT, 157
GRILSE, 71
Griseus, 178
Groundfish, 46
Grouper, 11, 24, 71, 175
Grumbler, 45
Grunniens, 161
Gurnard, 156
GURRY, 73
Gut cavity, 67
Guttatus, 72, 115

H & D, 80
H & G, 80

267

INDEX

H. stenolepsis, 75
HACCP, 83
Haddock, 54, 74, 172
Hake, 205
Halibut, 56, 57, 75
Haliotis, 2
Haliotis assimilis, 2
Haliotis corrugata, 2
Haliotis crackerodii, 2
Haliotis discus, 2
Haliotis kamtschatkana, 2
Haliotis rufescens, 2
Haliotis sorenseni, , 2
Haliotis spp., 2
Haliotis tuberculata, 2
Haliotis wallalensis, 2
HANDLING FRESH SEAFOODS, 76
HANDLING LIVE CRUSTACEANS, 79
HANDLING LIVE MOLLUSCS, 78
HANDLING SCALLOPS, 153
Handling shark, 159
Hard clam, 24
Hard shell clam, 19, 28
Hardhead, 21, 45
Harengus, 80
Harvestfish, 16
Hastata, 152
HATCHERY FISH, 80
Hazard Analysis of Critical Control Points, 83
HEAVY SALTED FISH, 147
Hen clam, 31
Henfish, 22, 100
Herring, 13, 15, 56, 80, 135, 172, 173, 204
Hind, 71, 72
Hinge width, 29
Hippoglossoides, 196
Hippoglossoides platessoides, 58, 59
Hippoglossoides robustus, 59
Hippoglossus, 75

Hippoglossus hippoglossus, 75
Hippos, 44, 86
Hippurus, 104
Histamine poisoning, 104
HISTAMINES, 81
Hogs, 30
Hoki, 206
Homarus americanus, 91
Homarus gammarus, 91
Hoolihan, 168
Hoplostethus atlanticus, 115
HORSE CLAM, 33
Horse crevally, 44
HORSE MACKEREL, 102, 194
HORSE MUSSELS, 110
Hot smoked salmon, 175
Hotels, 41
Humpback, 137, 143
Humper, 192
Huss, 161
Hypoxanthine, 68

Ice, 76, 83
Icelandic baby lobster, 99
ICSSL, 163
Ictalurus punctatus, 19
Ictalurus spp., 19
Ictiobus, 15
Ictiobus bubalus, 15
Ictiobus cyprinellus, 15
Ictiobus niger, 16
Idella, 18
IEF, 85
Iki-shime, 179
Ikura, 183
Illecebrosus, 181
Illex illecebrosus, 181
INCONNU, 83
Indian cure, 175
Inferior quality, 63, 65
Inferior size, 65
INK, 83
Inkfish, 45, 83
INSPECTION, 83

INDEX

Interstate Certified Shellfish Shippers List, 26, 163
Interstitialis, 73
Ionizing irradiation, 84
IPSWICH CLAMS, 26, 27
Irradians, 150
Irradiation, 4, 84
Irrideus, 141
Irroratus, 42
Islandica, 31
ISO-ELECTRIC FOCUSING, 85
Itajara, 72

J-cut, 54
Jack crevalle, 44, 86
Jack mackerel, 101, 102
Jack salmon, 142
JACKS, 5, 86, 198
Japanese eel, 52
Japanese littleneck, 32
Japanese oyster, 119
Japanese scallops, 150
Japanese yellowtail, 209
Japonica, 32
Japonicus, 52, 103
Jasus spp., 99
Jewfish, 11, 71, 72
Jimmies, 38
Jocu, 177
John Dory, 181
Johni, 100
JONAH CRAB, 42
Jordani, 58, 60
Jumbo, 35, 41, 87
Jumbo lump, 40

Kamaboko, 184
Kamtschatkana, 2
Kasu cod, 137
Katsuwonus pelamis, 195
Kazunoko-konbu, 80
Kegani crabmeat, 43
Kelp, 2

KENCH, 147
Keta, 137, 143
KILO, 88
Kilogram, 88
King crab, 38
KING MACKEREL, 103
King salmon, 141
King whiting, 88
KINGFISH, 88, 103
KINGKLIP, 88
KIPPER, 169, 173
KIPPERED FISH, 88
Kippered salmon, 175
Kisutch, 137, 142
Kitty Mitchell, 73
Knobbed whelk, 203
Kokanee, 144
Korean crab, 43
Korean variety crabmeat, 43
KOSHER, 88, 149
Kosher rules, 19
KRILL, 89
Kumamoto oysters, 119

L. princeps, 189
LABELING, 202
LACEY ACT, 90
LADYFISH, 90
Lake herring, 24
LAKE PERCH, 90
Lake trout, 190, 192
Lake Victoria perch, 111
Lalandei, 209
LAMPREY, 53
Lampris guttatus, 115
Lance, 168
LANE SNAPPER, 178
Langostinos, 100
Langoustine, 99
Lates niloticus, 111
Leiostomus xanthurus, 179
Lemon sole, 58, 59, 61
Leniusculus, 43
Lepidopsetta bilineata, 58, 61

INDEX

Lethostigma, 61
Leucichthys, 83
Levis, 132
Libertate, 189
Light meat tuna, 196
Limanda aspera, 62
Limanda ferruginea, 58, 62
LING, 90
Lingcod, 9, 34, 90
Lipomas spp., 183
Lisa, 107
LITTLE TUNA, 195
LITTLENECK, 26, 29
Littoralis, 88
Littorea, 124, 207
Littorina littorea, 124, 207
Liza, 107
Lobotes surinamensis, 190
Lobster, 12, 38, 79, 91
Lobster handling and storage, 97
LOBSTER, SPINY, 99
LOBSTERETTE, 99
Loco, 3
Loligo opalescens, 181
Loligo peallei, 180
Long finned sole, 60
Long finned squid, 180
Long rough dab, 59
Lophius americanus, 106
Lophius piscatorius, 106
Lopholatilus chamaelonticeps, 189
Lota, 16
Lota lota, 16
Lox, 174
Lucius, 124
Lump, 40
Lumpfish, 22, 100
Lumpsucker, 100
Lumpus, 100
Lurida, 120
Lutjanidae, 175
Lutjanus analis, 178
Lutjanus blackfordii, 176
Lutjanus buccanella, 177

Lutjanus campechanus, 176
Lutjanus cyanopterus, 177
Lutjanus griseus, 178
Lutjanus jocu, 177
Lutjanus mahagoni, 178
Lutjanus synagris, 178
Lutjanus vivanus, 178

Mackerel, 56, 82, 101, 173
Macrobrachium rosenbergii, 167
Macrocephalus, 34
Macropterus, 188
Macrouridae, 71
Macrozoarces americanus, 114
Macruronus novaezelandiae, 206
Mactromeris polynyma, 33
Maculatus, 50, 103
Magellanicus, 150
Magister, 41
Mahagoni, 178
MAHIMAHI, 49, 82, 104
Mahogany clams, 29
Mahogany quahog, 31
MAHOGANY SNAPPER, 178
Mako, 160
Malasol, 22
Mallotus villosus, 18
MANILA CLAM, 27, 32
MANTLE, 27, 180
MAP, 121
Marinated salmon, 71
Marinus, 53, 131
Market size grades for cod, 35
Market squid, 181
Marlins, 12
Masou, 137, 144
Masquinongy, 125
Masu salmon, 144
Maxima, 196
Medallions, 3
Medirostris, 183
Mediterranean mussel, 108
Medium red salmon, 142
Mediums, 41

INDEX

Melanogrammus aeglefinus, 74
Melanosis, 4, 105, 167
Melanosticus, 58, 61
MENHADEN, 105, 128
Menippe mercenaria, 41
Mentella, 131
Menticirrhus americanus, 88
Menticirrhus littoralis, 88
Menticirrhus saxatilis, 88
Menticirrhus spp., 88
Mercenaria, 28, 41
Mercenaria campechiensis, 29
Mercenaria mercenaria, 28
MERCURY, 105, 187
Merluccius australis, 206
Merluccius bilinearis, 206
Merluccius productus, 206
Merluza, 205
MERUS, 105
Metanephrops rubellas, 100
Microgadus proximus, 190
Microgadus tomcod, 190
Microlepis, 72
Micromesistius australis, 207
Micromesistius poutassou, 207
Micropogonias undulatus, 45
Microstomus pacificus, 58, 59
Milkfish, 8, 106
MILT, 70, 80, 106
MINCED CLAMS, 27, 31
Minced meat, 40
MISTY GROUPER, 72
Mock halibut, 196
Modified atmosphere packaging, 121
Modiolus spp., 110
Mola, 183
Mola mola, 183
MOLLUSC, 54, 106
Molva, 90
Molva molva, 90
MONKFISH, 106
Monoclonal antibodies, 85
Monopterygius, 9
Monterey squid, 181

Moonfish, 115
Mordax, 6
Morid cod, 34
Morio, 72
Morone chrysops, 204
Morone chrysops x saxatilis, 182
Morone saxatilis, 131, 182
Morwong, 188
Mossbunker, 105
MOULTING, 107
Moxostoma spp., 16
Mugil brabatus, 107
Mugil cephalus, 107
Mugil curema, 107
Mugil liza, 107
Mugil spp., 107
Mugil surmuletus, 107
Mugilidae, 70
Mullet, 16, 70, 107, 135
Mullet roe, 107
Mullidae, 70
MUSKELLUNGE, 125
MUSSEL, 46, 78, 108
Mustelus canis, 161
MUTTON SNAPPER, 178
Muttonfish, 114
Mya arenaria, 27
Mycteroperca bonaci, 72
Mycteroperca interstitialis, 73
Mycteroperca microlepis, 72
Mycteroperca phenax, 72
Mycteroperca venenosa, 73
Mykiss, 137, 141, 191
Mystacinus, 72
Mytilus edulis, 108
Mytilus galloprovincialis, 108
Myxosporidian, 58
Myxosporidian parasite, 205

N-3, 115
Namaycush, 192
NASSAU GROUPER, 72
National Fisheries Institute, 4

INDEX

National Shellfish Sanitation Program, 25, 47, 108, 116, 149
Nebulosus, 157
NECK, 27
NEEDLEFISH, 111
Nemadactylus macropterus, 188
Nephrops andamanicus, 100
Nephrops norvegicus, 99
Nerka, 137, 144
NET WEIGHTS, 111, 202
New Zealand whiting, 206
Niger, 16, 125
Nigritus, 73
NILE PERCH, 111
Niloticus, 111
Nitrites, 170
Niveatus, 72
NOBBING, 111
Nobilis, 11
NOMENCLATURE, 112
Northern abalone, 2
Northern anchovy, 6
Northern conch, 208
NORTHERN SHRIMP, 165
Norvegicus, 99
Norway lobster, 99
Noses, 66
Nothus, 157
Nova salmon, 174
Novaezelandiae, 206
NSSP, 26, 47, 108, 116, 131, 162, 164
Nuttall's cockle, 33
Nuttallii, 33

Obesus, 194
Ocean blowfish, 106
OCEAN CLAM, 31
Ocean crabs, 41
Ocean perch, 15, 130–132, 175
OCEAN POUT, 114

Ocean quahog, 31
Ocean whitefish, 189
Oceanicus, 37
Ocellata, 50, 130
Octopodidae, 114
Octopus, 23, 83, 114
Ocyurus chrysurus, 178
Odors, 66
Off odors, 49
Oglinum, 188
OLYMPIA OYSTER, 120
OMEGA-3, 115
Oncorhynchus aguabonita, 192
Oncorhynchus clarki, 192
Oncorhynchus gorbuscha, 137, 143
Oncorhynchus keta, 137, 143
Oncorhynchus kisutch, 137, 142
Oncorhynchus masou, 137, 144
Oncorhynchus mykiss, 137, 141, 191
Oncorhynchus nerka, 137, 144
Oncorhynchus tschawytscha, 137, 141
Onitis, 188
Ono, 202
OPAH, 115
Opalescens, 181
Opercularis, 151
Ophiodon elongatus, 90
Opisthonema libertate, 189
Opisthonema oglinum, 188
Orange roughy, 50, 88, 115
Ormer, 2
Osetra, 22
Osietr, 22
Osmeridae, 168
Ostrea chilensis, 120
Ostrea edulis, 120
Ostrea lurida, 120
Ostreidae, 115
Oxidation, 12, 129
Oxytetracycline (OTC), 7
Oysters, 8, 46, 78, 115, 173
Ozone, 46

INDEX

P. albigutta, 61
Pacifastacus leniusculus, 43
Pacific barracuda, 11
PACIFIC CALICO SCALLOP, 152
Pacific cod, 34
Pacific crayfish, 43
Pacific hair crab, 43
Pacific hake, 206
Pacific halibut, 75
PACIFIC LITTLENECK CLAM, 32
PACIFIC MACKEREL, 103
Pacific ocean perch, 132
PACIFIC OYSTER, 119
Pacific pompano, 16
Pacific razor clam, 32
Pacific red snapper, 132
Pacific Salmon Color Guide, 145
PACIFIC SARDINE, 148
Pacific saury, 111
Pacific snapper, 132
PACIFIC WHITING, 206
Pacificus, 58, 59
PACKAGING, 121
PADDLEFISH, 122
Paddlefish roe, 22
Painted mackerel, 102
Pale kings, 142
Palinurus spp., 99
Pallasii, 80
PALOMETA, 127
PAN FISH, 122
Pandalus borealis, 165
Pandalus platyceros, 166
Pandalus spp., 165
Panopea abrupta, 33
Panulirus argus, 99
Panulirus spp., 99
Paralichthys californicus, 75
Paralichthys dentatus, 58, 60
Paralichthys lethostigma, 61
Paralytic shellfish poisoning, 130
PARASITES, 5, 122, 149, 185
Parasitic worms, 12
Parophrys vetulus, 58, 59

Parvipinnis, 38
Pasteurized, 40
PASTEURIZING, 123
Patinopecten caurinus, 153
Patinopecten yessoensis, 150
Patula, 32
Paucispinis, 132
Peallei, 180
Pearls, 109
Pecten magellanicus, 150
Peelers, 38
Pelamis, 195
Penaeus aztecus, 166
Penaeus duorarum, 166
Penaeus setiferus, 166
Peprilus burti, 16
Peprilus simillimus, 16
Peprilus triacanthus, 16
Perca flavescens, 90
Periostracum, 31
PERIWINKLE, 78, 124, 207
PERMIT, 127
Perna canaliculus, 110
Petrale sole, 58, 60
Petromyzon marinus, 53
Phenax, 72
Philippinarum, 32
Phosphate, 48
PICKEREL, 125
PICKLE CURE, 147
Picowaving, 84
PIKE, 124
Pike-perch, 124
PIKEPERCH, 125
Pilchard, 80, 148
Pinbones, 14, 125
PINK, 143
Pink abalone, 2
Pink conch, 37
Pink salmon, 137
PINK SCALLOPS, 152
Pink shrimp, 166
Pintada, 102
Pinto abalone, 2
Piscatorius, 106

INDEX

Placopecten magellanicus, 150
Plaice, 58, 59
Platessoides, 58, 59
Platichthys stellatus, 58, 61
Platyceros, 166
Pleurogrammus azonus, 9
Pleurogrammus monopterygius, 9
Pogonias cromis, 50
Pogy, 105, 128
Pollachius, 125
Pollachius pollachius, 125
Pollachius virens, 125
Pollack, 125
Pollicipes, 10
Pollicipes pollicipes, 10
Pollicipes polymerus, 10
Pollicipes spp., 10
Pollock, 13, 55, 125, 126, 174
Polymerus, 10
Polynyma, 33
Polyodon spathula, 122
Polyprion spp., 208
Pomatomus saltatrix, 13
Pomfret, 16, 126
Pompano, 16, 126
PORGY, 128
Porpoise, 49
Potassium pyrophosphate, 48
Potassium tripolyphosphate, 48
Pounded lobsters, 94
Poutassou, 207
PRAWN, 128
Pressed caviar, 22
Primes, 41
Prionotus carolinus, 156
Probatocephalus, 161
Procambarus acutus, 43
Procambarus clarkii, 43
Productus, 206
Protothaca staminea, 32
Proximus, 190
Psetta maxima, 196
Psettichthys melanosticus, 58, 61
Pseudocyttus maculatus, 50
Pseudoharengus, 4

Pseudopleuronectes americanus, 58, 61
PSP, 32, 130
PUFFER, 128
Pulcher, 162
PUMPKIN, 27, 29
Pumpkinseed, 15
Punctatus, 19
Pups, 186

Quahaug, 28
Quahog, 24, 26, 27, 28
Quauhog, 28
Queen conch, 37
Queen scallop, 151
Quinault, 144
Quinqueradiata, 209

Rachycentron canadum, 33
Rainbow trout, 137, 190, 191
Raja spp., 167
Ranching, 8
Rancidity, 12, 129, 145
Ratpacking, 65, 129
Rays, 167
RAZOR CLAMS, 32
RECALL, 129
Recording thermometer, 188
Red, 144
Red abalone, 2
Red caviar, 22
RED CLAM, 33
Red crab, 38
RED GROUPER, 72
Red hake, 206, 207
Red herring, 169
RED HIND, 72
Red kings, 142
Red mullet, 107
Red snapper, 132, 175, 176
Red swamp crawfish, 43
RED TIDE, 116, 130, 149
Redfish, 11, 15, 130–132

INDEX

Regalis, 102, 157
Regular meat, 40
Reinhardtius hippoglossoides, 196
Retort pouches, 175
Retortable pouches, 172
REX SOLE, 60
Rhomboplites aurorubens, 178
Rigor mortis, 69, 131
RIPE FISH, 131
River herring, 4, 80
Robustus, 59
Rock bass, 11, 15
Rock cod, 132
Rock crab, 42
ROCK HIND, 72
Rock lobster, 99
Rock salmon, 161
ROCK SHRIMP, 134
Rock sole, 58, 61
ROCKFISH, 131, 132, 182
Roe, 38, 80, 100, 135, 149, 158, 188
Rollmops, 81
Romet-30, 7
Rosenbergii, 167
Rostrata, 52
Roughback, 61
Roundworm, 122, 149
Rubellas, 100
Rubida, 152
Rufescens, 2

S. ocellata, 130
Sablefish, 13, 16, 34, 55, 136, 174
Sagax, 148
Sailfish, 12
Saira, 111
Saithe, 125
Salar, 137, 139
Salmo aguabonita, 192
Salmo clarki, 192
Salmo gairdneri, 141, 191
Salmo irrideus, 141
Salmo salar, 137, 139

Salmo trutta, 192
Salmon, 6, 12, 55, 56, 137, 170
Salmon eggs, 22, 183
Salmon jerky, 175
SALMON TROUT, 144, 191
Salt cod, 145, 146
Salt fish, 145
Saltatrix, 13
Salted cod, 15
SALTED FISH, 145
Salvelinus alpinus, 23
Salvelinus fontinelis, 192
Salvelinus malma, 23
Salvelinus namaycush, 192
San Pedro squid, 181
Sand seatrout, 157
Sand sole, 58, 61
Sanddab, 58
Sandeel, 147, 168, 204
SANDLANCE, 147
Sandvein, 201
Sapidissima, 158
Sapidus, 38
Sarda chiliensis, 15
Sarda sarda, 15
Sarda, 15
Sardine, 80, 148, 204
Sardinella anchovia, 148
Sardinella aurita, 148
Sardinops sagax, 148
Sashimi, 136, 148, 193
SAUGER, 124
Saurus, 90, 111
Saxatilis, 88, 131, 182
Saxidomus giganteus, 32
Scad, 103
SCALES, 149
Scallops, 8, 12, 36, 48, 66, 149, 168
Scamp, 71, 72
Scampi, 99, 155
Schillerlocken, 161
Schrod, 156
Sciaenops ocellata, 130
SCIENTIFIC NAMES, 155

275

INDEX

Scomber japonicus, 103
Scomber scombrus, 101
Scomberesox saurus, 111
Scomberomorus cavalla, 103
Scomberomorus maculatus, 103
Scomberomorus regalis, 102
Scomberomorus sierra, 103
Scombridae, 101
Scombroid poisoning, 82
Scombrus, 101
Scooter, 11
Scrod, 35, 74, 155
Scungili, 37, 203
Scup, 15, 55, 122, 128
Sea bass, 71
Sea bream, 128
Sea clam, 31
SEA CUCUMBER, 156
Sea devil, 106
Sea eggs, 158
Sea perch, 71
Sea pout, 114
SEA ROBIN, 156
SEA SCALLOP, 150
Sea snail, 207
Sea squab, 128
SEA URCHIN, 158
SEATROUT, 11, 157
Sebastes alutus, 132
Sebastes fasciatus, 131
Sebastes goodei, 132
Sebastes levis, 132
Sebastes marinus, 131
Sebastes mentella, 131
Sebastes paucispinis, 132
Sebastes serriceps, 132
Sebastes spp., 132
Seelachs, 174
Selects, 117
Semi-boneless, 15
Semibright, 143
Semicossiphus pulcher, 162
Sepia, 45
Sepia spp., 45
Seriola dumerili, 5

Seriola lalandei, 209
Seriola quinqueradiata, 209
Serriceps, 132
Setiferus, 166
SEVICHE, 158
Sevruga, 22
Shad, 6, 158
Shark, 55, 56, 158, 187
Sharps, 29
Sheefish, 83
SHEEPHEAD, 162
SHEEPSHEAD, 161
SHELLFISH SANITATION, 162
Short finned squid, 181
Short shipment, 65
Short weight, 63, 65
Shrimp, 12, 128, 165
Shrimp aquaculture, 8
Sicyonia brevirostris, 134
SIERRA, 103
Siliqua patula, 32
SILK SNAPPER, 178
Silver dollar, 16
Silver eels, 52
Silver hake, 206
Silver mullet, 107
Silver perch, 161
Silver salmon, 142
SILVER SEATROUT, 157
Silverbright, 143
Silverside, 142, 168, 204
Simillimus, 16
Singing scallops, 152
Siphon, 25, 27
Siscowet, 192
SKATE, 167
SKIMMER, 27, 31
Skin, 67
Skipjack, 90, 195
Skipjack tuna, 193
Slabs, 41
Slime, 67
Slime sole, 59
Slimehead, 115
SLIPPER LIMPET, 168

INDEX

Smallmouth buffalo, 15
SMELL, 66
Smelt, 18, 168
Smoked chub, 13, 25
Smoked haddock, 173
Smoked salmon, 174
SMOKED SEAFOODS, 169
Smoked trout, 175
SMOLT, 175
Smooth dogfish, 161
Smooth oreo dory, 50
Snailfish, 100
Snails, 203
Snake, 11
Snapper, 13, 24, 74, 175
Snow cod, 126
Snow crab, 38
Snow scrod, 126
SNOWY GROUPER, 72
SOCKEYE, 144
Sockeye salmon, 137
Sodium, 48
Sodium bisulfite, 4
Sodium hexametaphosphate, 48
Sodium nitrite, 170
Sodium pyrophosphate, 48
Sodium tripolyphosphate, 48
SOFT SHELL CRABS, 38, 40
Soft shell crawfish, 44
SOFTSHELL CLAM, 26, 27
Softshell lobsters, 93
Solanderi, 202
Solea, 59
Solea solea, 59
Solea vulgaris, 59
Soles, 57
Solidissima, 31
Sooks, 38
Sorbic acid, 49
Sorenseni, 2
Sous-vide, 122
Southern blue whiting, 207
SOUTHERN FLOUNDER, 61
SPANISH MACKEREL, 103
SPANISH SARDINE, 148

Spathula, 122
Spearfish, 12
Special meat, 40
SPECKLED HIND, 72
Speckled seatrout, 157
Speckled trout, 157
SPENT, 179
Sphoeroides spp., 128
Sphyraena, 11
Sphyraena argentea, 11
Sphyraena barracuda, 11
Sphyraena sphyraena, 11
SPIKING, 179
Spiny dogfish, 161
SPINY SCALLOPS, 152
Spisula solidissima, 31
SPOT, 179
Spot shrimp, 166
SPOTTED SEATROUT, 157
Spotted tunny, 195
Spring herring, 4
Spring salmon, 141
Squalus acanthias, 161
Squeteague, 157
Squid, 3, 18, 23, 37, 55, 83, 179
Squirrel hake, 207
St. Peter's fish, 50, 181
St. Pierre, 50
Staminea, 32
Standards, 117
Starry flounder, 58, 61
STEAK, 181
STEAMER, 27
Steelhead, 137, 141, 144, 191
Stellatus, 61
Stenodus leucichthys, 83
Stenolepsis, 75
Stenotomus chrysops, 128
Sterlet, 22
Stickers, 42
Stimpson's surfclam, 33
Stizostedion canadense, 124
Stizostedion vitreum, 124
Stockfish, 145–147
Stomach, 27

INDEX

Stomias, 58
STONE CRAB, 41
Stones, 44
Strawberry grouper, 73
Striata, 11
Striatus, 72
STRIPED BASS, 11, 131, 182
Striped mullet, 107
Striped perch, 90
STRIPS, 27
Strombus gigas, 37
Strongylocentrotus spp., 158
STUFFED CLAMS, 27
Stuffies, 27
STURGEON, 182
Substitutions, 63, 64
Suckers, 15, 107
SUJIKO, 183
Sulfamerazine, 7
SULFITES, 4, 183
Sulphur dioxide, 183
Suminoe oyster, 120
Summer flounder, 60
Summer squid, 181
Sunfish, 115
SUNFISH, FRESHWATER, 183
SUNFISH, OCEAN, 183
Sunfishes, 15
Sunshine bass, 182
Superba, 89
SURF CLAMS, 27, 31
SURIMI, 184
Surinamensis, 190
Surmuletus, 107
Sushi, 123, 148, 158, 185, 193
Swordfish, 12, 55, 56, 186
Symmetricus, 102
Synagris, 178

Tags, 25, 33, 109, 164
Tailer, 13
Tanner crab, 38
Tapes philippinarum, 32
Tapeworm, 122, 149

TARAKIHI, 188
TARAMA, 188
TAUTOG, 188
Tautoga onitis, 188
Temperature, 68, 76
Tenpounder, 90
Tentacles, 180
Tenuis, 207
Tetracycline, 7
Thazard, 68, 102
Theragra chalcogrammus, 126
THERMOGRAPH, 188
THREAD HERRING, 188
Threaded abalone, 2
Thunnus alalunga, 194
Thunnus albacares, 196
Thunnus atlanticus, 194
Thunnus obesus, 194
Thunnus thynnus, 194
Thynnus, 194
Tilapia, 175, 181, 189
Tilapia spp., 189
TILEFISH, 189
Tinker, 101
TMA, 159
Tolerance levels, 7
TOMALLEY, 189
Tomcod, 34, 190
Tonguesole, 58
TOPNECK, 26, 27, 29
Toro, 44
Trachinotus carolinus, 127
Trachinotus falcatus, 127
Trachinotus goodei, 127
Trachinotus spp., 126
Trachurus, 102
Trachurus symmetricus, 102
Trachurus trachurus, 102
Transmontanus, 183
Trap squid, 180
Treefish, 132
Trematode parasite, 51, 130
Tresus spp., 33
Triacanthus, 16
TRIGGERFISH, 190

INDEX

Trimethylamine, 159
TRIPLETAIL, 190
Triploid, 118
TRIPLOIDY, 190
TROPICAL SHRIMP, 166
TROUT, 175
TROUT, FRESHWATER, 190
True cod, 34
Trutta, 192
Tschawytscha, 137, 141
Tuberculata, 2
Tullibee, 24
Tullie, 141
TUNA, 82, 193
Turbot, 8, 57, 58, 59, 196
TURBOT GREENLAND, 196
Tusk, 45
Tyee, 141
Tyrannus, 105

Ultraviolet, 47
ULUA, 198
UNDERUTILIZED SPECIES, 198
Undulatus, 45
Unique radiolytic products, 84
Univalves, 106
Urea, 159
Uric acid, 13
Urophycis chuss, 207
Urophycis tenuis, 207
URPs, 84

V-cut, 35, 54
Vacuum packed foods, 6
Vacuum packing, 172
Vaillanti, 19
VEIN, 201
VENDACE, 201
Venenosa, 73
Venerupis japonica, 32
Venerupis philippinarum, 32
VERMILION SNAPPER, 178

Vetulus, 58, 59
Vibrio, 162
Villosus, 18
Virens, 125
Virginica, 119
Viruses, 162
Vitreum, 124
Vivanus, 178
Viviparous blenny, 114
Viviparus, 114
Vulgaris, 59

WAHOO, 202
Wallalensis, 2
WALLEYE, 124
Walleye pike, 124
Walleye pollock, 126
Wampum, 29
Warsaw, 71, 73
WATERMARK, 202
Weakfish, 157
WEATHERVANE SCALLOP, 153
WEIGHTS, 202
Whales, 35, 41
Wheels, 186
WHELK, 37, 124, 203, 208
White abalone, 2
White amur, 18
WHITE BASS, 11, 182, 204
White chinooks, 142
WHITE HAKE, 207
White herring, 4
White meat tuna, 194
White mullet, 107
White river crawfish, 43
WHITE SEABASS, 11
White seatrout, 157
White shrimp, 166
WHITE STURGEON, 183
Whitebait, 7, 148, 168, 204
Whitefish, 83, 175, 205
WHITEFISH, LAKE, 205
Whiting, 55, 88, 122, 205
WHITING, BLUE, 207

279

INDEX

WHO, 36
Wing, 168, 180
WINKLE, 124, 207
Winter flounder, 58, 61
Witch, 60
Witch flounder, 60
Witch sole, 60
Wolffish, 21
World Health Organization, 36
Wrasse, 162, 188
WRECKFISH, 208

Xanthurus, 179
Xiphius gladius, 186

Yellow eels, 52
Yellow perch, 90
Yellow pickerel, 124
Yellow pike, 124
Yellow pout, 114
Yellow walleye, 124
YELLOWEDGE GROUPER, 73
YELLOWFIN, 196
YELLOWFIN GROUPER, 73
YELLOWFIN SOLE, 62
YELLOWMOUTH GROUPER, 73
YELLOWTAIL, 5, 8, 58, 209
YELLOWTAIL FLOUNDER, 62
YELLOWTAIL SNAPPER, 178
Yessoensis, 150
YIELD, 209

Zachirus, 60
Zenopsis ocellata, 50
Zeus faber, 50
Zoarces viviparus, 114